CAMBRIDGE COUNTY GEOGRAPHIES

General Editor: F. H. H. GUILLEMARD, M.A., M.D.

THE

EAST RIDING

OF

YORKSHIRE

Cambridge County Geographies

THE
EAST RIDING
OF
YORKSHIRE
(WITH YORK)

by

BERNARD HOBSON, M.Sc., F.G.S.

Late Lecturer in Geology in the Victoria University of Manchester

Author of *The West Riding of Yorkshire*

With Maps, Diagrams, and Illustrations

Cambridge:

At the University Press

1924

CAMBRIDGE UNIVERSITY PRESS

Cambridge, New York, Melbourne, Madrid, Cape Town,
Singapore, São Paulo, Delhi, Mexico City

Cambridge University Press
The Edinburgh Building, Cambridge CB2 8RU, UK

Published in the United States of America by Cambridge University Press, New York

www.cambridge.org
Information on this title: www.cambridge.org/9781107690356

First published 1924
First paperback edition 2013

A catalogue record for this publication is available from the British Library

ISBN 978-1-107-69035-6 Paperback

PREFACE

THE writer is indebted to the Director of the Geological Survey for permission to reproduce the Drift Geological Map, Rainfall Map, and the Section from Filey Brigg to Speeton Cliff; to the Director of the Ordnance Survey for statistics of areas; to the Meteorological Office for rainfall data. Mr J. R. Procter has kindly supplied information on administration and fisheries, and with Messrs H. Sheffield, T. B. Todd, H. S. Powell, P. J. Spalding, and R. B. Dunwoody on land drainage and inland navigation. Prof. A. Mawer has given help with place names, Mr S. H. Smith with zoological information, Mr A. Whitehead, of the Hull Chamber of Commerce and Shipping, with shipping data, and Mr Godfrey Bingley has given the use of his photographic negatives. Others who have supplied illustrations are mentioned subsequently. The publishers of the *Dictionary of National Biography* have given permission to quote from that work the details (with a few exceptions) given in the Roll of Honour. To all those mentioned and to others who have kindly supplied information and to Dr Guillemard, the General Editor of the Series, the writer desires to express his cordial thanks. He will be obliged by information of any errors.

B. H.

20 HALLAMGATE ROAD
SHEFFIELD
1924

CONTENTS

		PAGE
1.	County and Shire	1
2.	General Characteristics. Position and Natural Conditions	2
3.	Size. Shape. Boundaries	4
4.	Surface and General Features	6
5.	Watersheds. Rivers. Lakes	11
6.	Geology	18
7.	Natural History	32
8.	Peregrination of the Coast	40
9.	Coastal Gains and Losses	48
10.	Climate	54
11.	People—Race, Language, Settlements, Population .	59
12.	Agriculture	65
13.	Industries	68
14.	Minerals	71
15.	Fisheries	75
16.	Shipping and Trade	78
17.	History	84

PAGE

18. Antiquities—(*a*) Prehistoric 91

19. Antiquities—(*b*) Roman 96

20. Antiquities—(*c*) Anglo-Saxon 101

21. Architecture—(*a*) Ecclesiastical 104

22. Architecture—(*b*) Military 125

23. Architecture—(*c*) Domestic 131

24. Communications—Past and Present—Roads, Canals,
    Railways 137

25. Administration and Divisions—Ancient and Modern . 145

26. The Roll of Honour 148

27. The Chief Towns and Villages of the East Riding . 157

ILLUSTRATIONS

	PAGE
Dry Chalk Valley, Thixendale	15
Hornsea Mere	17
Section from Filey Brigg to Speeton Cliff	23
Diagram of Drainage of the Vale of Pickering . . .	24
Section of Buried Cliff at Sewerby near Bridlington Quay .	27
Geological Map of the East Riding showing Glacial Drift .	30
Great Crested Grebe	36
Nest of Little Grebe	37
Ringed Plover	38
Egg Collecting, Bempton Cliffs	39
Doodle or Cove on North side of Carr Naze . . .	41
Filey Brig	42
Weathered Boulder Clay overlying Corallian Rocks, Carr Naze	43
East sides of Chatterthrow, Little and Great Thornwick Bays, Flamborough Head	45
Map of Average Annual Rainfall of the East Riding .	58
King George Dock, Hull	82
Flint arrow-head, Bridlington	92
Cinerary urn, Goodmanham	94
Sword and scabbard, Grimthorpe	96
Map of Roman Roads of the East Riding . . .	99

	PAGE
Pre-Conquest Cross, North Frodingham	103
Norman Chancel Arch, Goodmanham Church	107
Weaverthorpe Church	108
Rood Screen, Flamborough Church	109
Norman Font from Hutton Cranswick, now in York	110
Filey Church	111
Patrington Church	112
Easter Sepulchre, Patrington Church	113
Monuments in Hilton Chapel, Swine Church	116
Gatehouse and base of Cross, Kirkham Priory	117
Oriel Window, Watton Priory	118
York Minster, West Front	120
York Minster, the Choir	121
Beverley Minster, West Front	122
Percy Tomb, Beverley Minster	123
Howden Church	124
The Bayle Gate, Bridlington Priory	126
Wressell Castle	128
Hollar's View of Hull in 1640	129
Micklegate Bar, York	131
Half-timbered Houses, York	133
Gatehouse, Burton Agnes Hall	135
Jacobean Doorway, King's Manor, York	136
A Yorkshire " keel " on Beverley Beck	140
Andrew Marvell	153
William Wilberforce	156
Saturday Market Place, Beverley	158

PAGE

Church and village of Bishop Burton   159

Old Kilnsea Cross   163

Humber, Prince's, and Queen's Docks, Hull   .   .   . 165

Diagrams   171

The illustrations on pp. 41, 42, 43, 45, 109, 111, 112, 113, 120, 122, 123, 124, 126, 133 are from photographs kindly supplied by Mr Godfrey Bingley; that on p. 17 from a photograph supplied by Mr R. Fortune; the section on p. 23, the Drift map, p. 30, and the Rainfall map, p. 58, are reproduced from Geological Survey publications; the diagram p. 24 is by Dr F. R. Cowper Reed; the section on p. 27 is reproduced by permission of Mr G.W.Lamplugh; the illustrations on pp. 36, 38 are from photographs by Mr K. J. A. Davis; that on p. 37 is from a photograph by Mr T. L. Smith; that on p. 39 was supplied by Lieut. A. H. Robinson; that on p. 82 is from a photograph supplied by Mr M. Barnard; the illustrations on pp. 92, 94, 96 were supplied by the British Museum; that on p. 103 was supplied by Mr W. G. Collingwood; the illustrations on pp. 15, 107, 116, 117, 118, 128, 135, 140, 158, 159 are from photographs by Mr B. Hobson; that on p. 108 is from a photograph supplied by the Rev. A. G. Braund; those on pp. 110, 136, 163 were supplied by Dr W. E. Collinge (Yorkshire Philosophical Society); that on p. 121 is from a photograph supplied by F. Frith & Co. Ltd.; that on p. 130 was supplied by A. Brown & Sons, Ltd.; that on p. 153 by Mr W. H. Bagguley; that on p. 156 by Airco Aerials Ltd.

1. County and Shire.

The division of England into counties and shires did not arise, as might be thought, from the action of some supreme authority portioning off a hitherto undivided country on a thought-out plan. On the contrary, the kingdom is the result of the union or aggregation of previously separate and, in many cases, independent areas. Some of these areas, such as Kent, Sussex, Essex, Middlesex and Surrey were ancient kingdoms, while others, such as Norfolk and Suffolk, are divisions of ancient kingdoms, in their case of East Anglia.

Yorkshire represents the ancient Anglian kingdom of Deira, though it was frequently united with Bernicia in the larger kingdom of Northumbria. It was recognised as a shire before the Norman Conquest, and appears in Domesday Book (1086) as Eurvicescire or Euruicscire.

Some derive the word shire from the Anglo-Saxon *sciran*, to shear, indicating a part shorn off from a larger division, but more recent authorities hold it to be from *scīr*, meaning office, charge, or administration; hence a district governed by one or more officials. These officials, before the Norman conquest, were the sheriff (shire-reeve, from Anglo-Saxon *scīr-gerefa*) and an ealdorman or earl. The word county was introduced by the Normans, and is derived from the Anglo-French *conté*, from Old French *conte* or *comte* = a count. A county is the district originally governed by an earl or count, and the word is now used as an equivalent to shire. The origin of the name York-shire is almost self-evident. It is the shire of which York is the capital.

The Celtic name of York was Eburac (of unknown meaning) which the Romans Latinized as Eburăcum or Eborăcum. The Angles called it Eoforwic, "the lair of the wild boar," Eoforwicceaster, Everwic. The Danes called it Ioforvik (the initial I having the sound of Y) later contracted to Iorvik, whence Yorick and York. In Domesday Book the name is Euruic.

Even before the Norman conquest Yorkshire was divided into three parts (conveniently meeting at York), which we now call ridings. The word riding is derived from an Anglo-Saxon adaptation of Old Norse, meaning "a third part," whence the Early Middle English *trithing* or *triding*. The initial *t* afterwards became absorbed in the final *th* or *t* of the words North, East and West.

2. General Characteristics. Position and Natural Conditions.

Yorkshire, the largest of English counties in area and surpassed only by Lancashire and London in population, is situated in the north-east of England and stretches from the North Sea to within eight miles of the Irish Sea. Its coast line extends from the estuary of the Tees to that of the Humber. The three ridings into which this great county is divided afford an interesting contrast. The North Riding partakes in some respects of the features of the other two, while the East and West Ridings are almost the antitheses one of the other.

In the North Riding we have two groups of hills separated by a great plain; in the West Riding lofty hills on the west and a plain on the east; in the East Riding hills of no great elevation in the centre, flanked by plains to east, west, and north, and a great estuary on the south.

Unlike the West Riding, the East Riding has no un-inhabited moorland hills. The most characteristic feature of the riding is the hill range of the Chalk Wolds, which sweeps in a crescentic curve from Hessle on the Humber estuary to terminate in the magnificent sea-cliffs of Flamborough Head, one of the finest promontories of England. Yet the loftiest summit of this hill range barely exceeds 800 feet in height and forms so slight an obstacle to travel that its highest point is crossed by a high road of prehistoric antiquity. The Wolds turn their escarpment to the Vale of York on the west and the Vale of Pickering on the north, their gentle dip-slope to the low-lying plain of Holderness on the east. These plains afford ample scope for agriculture (68 per cent. arable to 32 per cent. pasture), which even extends on to the Chalk Wolds.

Agriculture is as characteristic of the East Riding as manufacturing and mining industry is of its western neighbour. The East Riding is a maritime county, possessing in the city of Hull on the Humber estuary the third port of the United Kingdom, and, on the sea-coast, such popular watering places as Filey with its famous Brig, Bridlington, Hornsea with its mere, and Withernsea.

3. Size. Shape. Boundaries.

Though for convenience in administration Yorkshire is divided into three Administrative Counties, eleven county boroughs (of which the North and East Ridings contain only one each) and the city and county of the city of York, the ancient county deserves consideration as a whole.

The following table shows the area of its main divisions:

Administrative Counties and Associated County Boroughs	Inland Water, Acres	Land and Inland Water, Acres	Square Miles	Square Root of Area, Miles	Percentage of Total Area	Proportion of the Divisions
West Riding	13,955	1,773,529	2,771	52·6	45·6	2·4
North Riding	4,853	1,362,058	2,128	46·1	35	1·8
East Riding	1,852	750,115	1,172	34·2	19·3	1
York City	89	3,730	5·8	2·4	0·1	·005
	20,749	3,889,432	6,077	78	100	

The square root of area shows the length of the side of a square which would have the area in square miles given in the preceding column.

Yorkshire is more than four-fifths of the size of Wales, more than twice the size of Lincolnshire, the next largest county, and nearly forty times the size of Rutland.

The East Riding (including Hull) has strictly speaking a larger area than that shown in the preceding table. It consists (in acres) of Land 748,263, Inland Water 1852, Foreshore 18,828, Tidal Water (other than the Humber) 4179, Tidal Water of the Humber 23,311, Total 796,433 acres or 1244 square miles. For comparison with inland

counties we may take its area as 1172 square miles. The
county which the East Riding most nearly resembles in
area is Staffordshire but the East Riding is the larger by
13¾ square miles.

The greatest length of the riding from the bend of the
Derwent at Marishes, near Malton, to Kilnsea is 52 miles.
The longest line at right angles to the length is from the
Hertford River, near Seamer Carr, to the bend of the
Ouse below Goole, 40 miles. The outline of the riding is
most simply described as pear-shaped or kite-shaped, the
stem of the pear being situated at Spurn Head; it is an
irregular pentagon, the angles being at Filey, Marishes,
Cawood, Goole, and Spurn.

The riding is bounded on the north and north-west by
the North Riding from the Wyke, near Filey, to York,
then for a short distance by the municipal boundary of
York; on the south-west by the West Riding from York
to the confluence of the Ouse and Trent; on the south
by an imaginary line running approximately down the
middle of the Humber, the other side of the line being
tidal water of Lincolnshire; on the east by the North Sea.
Until April 1, 1889, part of the township of Filey was in
the North Riding; on that date it was wholly included
in the East Riding. The present boundary of the East
Riding starts at the Wyke in the cliffs N.W. of Filey
and runs almost due south for a mile, turns N.W. and N.
for half a mile and then follows the Hertford River and
the old course of that river for 7½ miles to its junction
with the Derwent. It follows the old course of the
Derwent for 11½ miles, when it diverges to the north of the
existing river for a boundary length of 1½ miles (probably

the old course), rejoining the Derwent near Marishes, thence following the Derwent for 21½ miles to Stamford Bridge. From Stamford Bridge it follows the line of the high road (originally Roman or pre-Roman) for 6¼ miles to the municipal boundary of York, which it follows for 1¼ miles to the Ouse at Fulford, with whose windings it coincides for 41 miles to the junction of the Ouse and Trent, opposite Faxfleet. The total length of boundary from the Wyke to Faxfleet is 92 miles. From the confluence of the Ouse and Trent the boundary (generally speaking but not always) follows the middle line of the Humber at low water for 38¼ miles, when it turns at right angles for 2 miles to the south end of Spurn Head. The length of the north bank of the Humber measured at high-water mark from the confluence of Ouse and Trent to the south end of Spurn Head is 43½ miles. The length of the North Sea coast of the riding from the Wyke near Filey to the south end of Spurn Head is 60 miles.

4. Surface and General Features.

Geographically the East Riding may be divided into seven regions (of which the first three are the most important), viz.: 1. Holderness, 2. The Yorkshire Wolds, 3. The Vale of York, 4. The Howardian Hills, 5. The Vale of Pickering, 6. The North Sea coast, 7. The North coast of the Humber estuary.

1. Holderness. This is the ness or promontory of Holder (of uncertain etymology). The term Holderness is here used in a wide geographical sense for the low-lying plain east and south of the Wolds. In shape Holderness

reproduces, on a smaller scale, that of the riding as a whole. Its sea-coast extends for nearly 41 miles from the buried Chalk cliff at Sewerby, north of Bridlington, to the south end of Spurn Head, and its Humber estuarine coast for $32\frac{1}{2}$ miles from Spurn to Hessle Cliffs, where the Chalk Wolds reach the Humber. Its inland boundary cannot be so clearly defined, but it approximately follows a buried sea-cliff of Chalk through Bridlington, Driffield, Beverley, and Cottingham to Hessle.

Holderness consists exclusively of superficial deposits—glacial boulder-clay, sand and gravel, and more recent river alluvium and lacustrine deposits. These are underlain by the Chalk, which crops out in the Wolds to the west and north. The whole of Holderness is a plain of very slight elevation. Most of the region is only from 10 to 30 feet above sea level. It rises to 100 feet at the foot of the Wolds and it also rises towards the sea-coast. Its surface is undulating except on the alluvial flats of the River Hull. Although it has few woods the numerous hedgerow-trees prevent it from seeming woodless.

The scenery of Holderness is naturally very tame, and the land has been so well drained and cultivated that, except in a few places, it has lost its original character. Before the advent of man it was characterised by numerous meres, of which only that of Hornsea remains, and by isolated hills of boulder-clay, which were separated one from another by marshes impassable either on foot or in boats.

2. The Wolds. Derived from the Anglo-Saxon *weald* which was commonly used in the sense of waste ground, the Wolds were probably never much wooded. They

form the most characteristic feature of the East Riding and
consist of a range of escarpment hills of Chalk, in this
case using the term escarpment in the wide sense, for the
elevated country formed by the outcrop of certain strata.
They run N.N.W. from North Ferriby on the Humber
to Acklam, thence with a much indented margin N.E. to
West Heslerton, thence E. by N. to Muston, thence S.E.
by E. to Flamborough Head. They form a crescent with
its convex side turned to the west and north, and measuring
about 56 miles along its convexity, ignoring all indentations.
The beds of Chalk are gently inclined east-south-eastward
with many local variations and they have been abruptly
truncated by subaerial erosion on the west and north. On
the west, as far north as Huggate, they can hardly be said
to form an escarpment, in the restricted sense of a cliff.
It is only along the northern border, from Knapton
eastward, that there is a continuous escarpment-ridge.
On their eastern and southern sides they form a gentle
dip-slope. This is by no means a uniformly inclined plane
surface but is sculptured by streams into hills and valleys,
the latter generally dry (see Chapter 5). The Chalk is
continuous under the bed of the Humber and forms the
Lincolnshire Wolds and similar escarpment hills of Chalk
form the North and South Downs. The general elevation
of the Wolds is from 400 to 600 feet, and their greatest
height is 808 feet at Garrowby Top on the main road
from Stamford Bridge to Driffield, ¾ mile E.N.E. of
Wilton Beacon (785 feet). Their area between 250 and
500 feet has been measured as 205 square miles and that
over 500 feet as 51·2 square miles. The area of Chalk
uncovered by later deposits is 294 square miles. The

width of the Wolds varies from 4 to as much as 12 miles, and they are traversed by several depressions which in mountains would be called passes.

In fine weather the Wolds have a considerable attraction arising from the clearness and purity of the air, the brilliant sunlight reflected from the white roads, the gently swelling and undulating surface, and the wide expanse of open country with scarcely a house or a tree in sight. Here and there, however, in the valleys trees and coppices occur and secluded villages nestle among them.

3. The Vale of York, sometimes called the Levels, forms the great central plain of Yorkshire and extends into all three ridings. Its surface sometimes consists of boulder-clay but chiefly of sand, gravel, warp, lacustrine clay, and river alluvium. The width of the vale on the north between Ingleby Arncliffe and Middleton Tyas is 13 miles; on the south between North Ferriby and Pontefract 30 miles. Its length from Great Smeaton to Snaith is 52 miles. Of this great area only the south-eastern portion belongs to the East Riding. It measures 21 miles from N. to S. between Skirpenbeck and Blacktoft, by 19 miles from E. to W. between Market Weighton and Cawood. This plain gives the impression of being a dead level; much of it is less than 25 feet above sea level and the rest hardly ever exceeds 50 feet. The hill of Holme on Spalding Moor, consisting chiefly of Keuper Marl, rises to 150 feet, like an island above the plain, and the view from its summit is very attractive. In the Vale of York "vast commons have been enclosed and cultivated, and dreary wastes full of swamps, which could not be crossed without danger, are now covered with well-built farm houses."

These are generally built of red brick and roofed with tiles of the same colour. The fields are divided by quickset hedges instead of the dry stone walls of the Millstone Grit country near Sheffield.

4. The Howardian Hills are called after the palatial seat of Castle Howard in the North Riding, and extend from Gilling Gap, which separates them from the Hambleton Hills, S.E. by E. for 19 miles to the foot of the Wolds between Settrington and Acklam. They are divided into two unequal portions by the deep and winding gorge of the Derwent. Only the portion on its left (S.E.) bank is in the East Riding. It measures only four or five miles from N.W. to S.E., but six miles at right angles to that direction. Its greatest height is about 400 feet at the foot of the Wolds.

5. The Vale of Pickering is named from the town in the North Riding. It is bounded by the Tabular Hills on the north, part of the Hambleton Hills on the west, the Howardian Hills and the Wolds on the south, and the Filey ridge of boulder-clay on the east. It measures 28 miles from E. to W., by a maximum of 10 miles from N. to S., its area being 140 square miles, and its soil is of sand and gravel, alluvium and lacustrine clay, which form a level plain from about 60 to 100 feet above sea level. It is the naturally drained basin of a great glacial extra-morainic lake and even at the present day, in spite of artificial drains, the land has a marshy water-logged appearance. Only that portion of the vale lying to south of the Hertford and Derwent rivers and measuring 21 miles in length on the curve by $3\frac{1}{2}$ miles in maximum width, is in the East Riding. Its floor has an area of 38 square miles, covered by Post-Glacial deposits.

6. The North Sea Coast. This is described in Chapter 8.

7. The North Coast of the Humber estuary consists almost entirely of alluvium and estuarine warp (the light chocolate-coloured sediment deposited by the tidal water of the Humber and its tributaries). It is a very low-lying area and "with the exception of a short distance at Paull, where the Humber cuts into the morainic hill [40 feet high] there, the whole of the land bordering the Humber between Hull and Kilnsea is reclaimed land and the tidal water is kept from overflowing the country by an artificial bank." (See Chapter 9, Coastal Gains and Losses.) Between Hull and Hessle low-lying alluvium prevails. Between Hessle Cliff and North Ferriby the Wolds, overlain by boulder-clay, are only separated from the Humber by a narrow strip of gravel or silt, but to the west of them the low-lying alluvial deposits of the Vale of York form the north shore of the estuary.

5. Watersheds. Rivers. Lakes.

The only watershed of much importance in the riding is that of the Wolds, which separate the waters flowing to the Derwent and Foulness from those of the Hull and other streams of Holderness.

The Ouse, with its tributaries, is the chief river of Yorkshire. It drains an area of 4187 square miles of which 4035 square miles are in Yorkshire (66·4 per cent. of the area of the county). It is formed by the junction ($2\frac{1}{4}$ miles E.S.E. of Boroughbridge) of the Swale, 66 miles long, which rises within two miles of the Eden, and the Ure, 61 miles long, rising on the east flank of a continuous valley, connecting at 1194 feet with that of the Eden. The tributaries of the Ouse in the East Riding are the Stillingfleet Beck, 11 miles long, and the Derwent.

At Trent Falls, after a course of 60 miles 6 furlongs, the Ouse joins the Trent to form the Humber. As the whole course of the Ouse is through a low-lying, almost level alluvial plain its flow is sluggish (750 yards per hour at summer level at York) and the scenery along its course is tame, but the river is impressive from its width and the volume of water it discharges which at York is 306,417 million gallons (over $\frac{1}{3}$ cubic mile) per annum. It annually transports past York in suspension and solution 300,000 tons of solid matter.

The Derwent rises in the North Riding on the east flank of Lilla Rigg in the Moorland Hills. It then flows due south through the deep and wooded gorge of Langdale and, after passing Hackness, enters the beautiful gorge of Forge Valley, from which it emerges into the level plain of the Vale of Pickering. It is there joined by the Hertford River on the left, turns westward, and begins to form the boundary of the East Riding. Four miles above New Malton it is joined by the Rye on the right (see sketch-map, p. 24, stage B). Nearly three miles below Malton the Derwent, now less than 50 feet above sea level, enters a beautiful winding wooded gorge, $5\frac{1}{2}$ miles long and 200 feet deep, which cuts through the Howardian Hills and is followed by the railway between Crambe and Huttons Ambo station. In this gorge are the ruins of Kirkham Priory (Abbey). Emerging into the Vale of York, the Derwent flows past Stamford Bridge and joins the Ouse near Barmby-on-the-Marsh. At Loftsome bridge below Wressell the

Derwent is 84 feet wide. The length of the Derwent is about 68 miles and its basin has an area of 757 square miles, of which 248 are in the East Riding. The remarkable history of the evolution of the Derwent is described in chapter 6 on Geology, p. 28. The left bank tributaries of the Ouse, other than the Derwent, drain 82 square miles in the riding.

The river Foulness (*pron*. Foonay) rises east of Market Weighton and after an almost semi-circular course of 17 miles ends in the Market Weighton Canal, which after 4¾ miles joins the Humber. Its catchment basin, including Mires Beck, is 133 square miles.

The River Hull rises near Great Driffield in strong springs issuing from the Chalk at Emswell and Little Driffield and in the Eastburn Beck. The Hull, confined between embankments and fed by becks and artificial drains, flows southward, bordered by wide stretches of alluvium, often peaty, leaving Beverley a mile to the west, and joins the Humber in the city of Kingston-upon-Hull. Its length, including the Emswell Beck, is at least 27 miles and its drainage area 364 square miles.

The Humber was originally the lower course of the Aire, but is now the great estuary of the united Ouse and Trent, which drain an area of 8204 miles. In addition, the north bank tributaries of the Humber drain 675 and its southern tributaries 428 square miles. The length of the estuary is 31 miles, its width at high water at Trent Falls 5 furlongs; at South Ferriby 2¼ miles; at Hull 1 mile 6½ furlongs; east of Skeffling 7½ miles; at the Spurn 4 miles 6 furlongs. Its area at high water is about 72,641 acres or 113½ square miles. The course of the Humber is deflected by the hard deposit of Glacial gravel at Paull, which imparts to it a south easterly direction, but in pre-glacial times the Humber flowed eastwards towards the site of Withernsea, as is shown by borings through the drift deposits. Between Whitton and Hessle the Humber cuts through three successive escarpments, that of the Lias at Whitton and of the Inferior Oolite at Winteringham, each over 100 feet high, and that of the Chalk between North and South Ferriby, over 200 feet in height. It is clear that the Aire-Humber began to flow before these escarpments existed when the Lias, Oolites, and Chalk extended far to the west of their present outcrops and the river flowed over the Chalk at a level several hundred feet higher than its present one.

The streams flowing directly into the North Sea such as the Speeton Beck, the Gipsey Race at Bridlington, and the Stream Dike at Hornsea are of small importance. According to Dr H. R. Mill they drain 171 square miles.

There remain to be considered the streams of the Chalk Wolds. They are very few and small, and yet the annual rainfall on the Wolds is from 5 to 7½ inches greater than in the low-lying plain, and the Wolds, especially between the Vale of Pickering and Driffield, are furrowed by a most beautifully ramifying series of valleys, from 30 to 200 feet deep. These valleys agree in having bottoms flat in *trans-verse* section consisting of gravel composed entirely of chalk and flints. The main valley of the Wolds is called the Great Wold Valley or Gipsey Race Valley. It begins at the western edge of the Wolds, at Wharram-le-Street and, in its course of 24 miles, includes Kirby Grindalythe, Weaver-thorpe, North Burton, and Rudston, ending at Bridlington

Quay. The valleys of the Wolds have undoubtedly been excavated by running water, though they are now usually dry. To account for this anomaly it is necessary to consider the behaviour of chalk. It is an exceedingly porous rock, so that of the rain which falls upon it, that which does not evaporate soon sinks in and continues to percolate down-

Tributary dry Chalk valley joining Thixendale counter
to its slope near Burdale station

wards until it reaches the plane of saturation or water-table, below which the chalk is saturated with water. This is not a true plane but has an undulating surface, rising higher under hills. So long as the saturation-plane does not reach the surface of the ground no water will flow at the surface but, as soon as the saturation-plane reaches the surface, springs will burst out. In this way the celebrated Gipsey springs (the G is hard, as in gimlet) are produced. They are intermittent springs like the bournes, nailbournes or winterbournes of southern England, e.g. Croydon. The best known spring is called the Gipsey Race and usually appears after a wet season in winter or early spring, being sometimes inactive for three consecutive years. The source of the Gipsey Race is in the Great Wold Valley near Wold Newton, east of Foxholes. The water does not rise in a body but oozes and trickles through the grass, where the ground is not broken, and flows away in a channel twelve feet wide and three feet deep, reaching the sea in Bridlington Harbour, after a course of 12½ miles. The volume of water is often very small. On April 18, 1922, at Boynton it was only 10 feet wide and about 6 inches deep.

To return to the difficulty in accounting for the ramifying Wold valleys, now almost always dry, the accepted view at present is that they were excavated by streams during the Great Ice Age, when the rainfall was greater and the Chalk was impervious, owing to its pores being closed by the freezing of water in them. Where impermeable rocks, such as clay, underlie the Chalk, as is usually the case along the Wold escarpment, springs occur at the contact. This circumstance "has influenced the selection of the sites of villages" in the Vale of Pickering (and elsewhere) such as Hunmanby, Folkton, Flixton, Staxton, Ganton, Sherburn, Heslerton, Winteringham, Thorpe Basset, and Rillington, which lie at the foot of the Chalk escarpment, where "springs of beautiful calcareous water burst out."

The only lake in the riding, and the largest in Yorkshire, is Hornsea Mere which has an area of 327·5 acres. It measures nearly a mile and a half in length by five furlongs in breadth. Its surface is only about 12 feet above sea-level and its depth is not more than 12 feet. Its margin is wooded. Reeds 10 to 14 feet high form in places a jungle. They are accompanied by bulrushes, sedges, and water-lilies. The

Hornsea Mere

lake is a moraine mere and was formerly much larger. "Even without allowing for the part lost by marine denudation, the old lake would measure two-thirds of a mile wide by at least 3 miles long." Formerly moraine meres were numerous in the Holderness, as is indicated by the name Mere attached to a farm, a field, or even to a hill. Many more besides those mentioned in Chapter 6 (Geology, p. 30) existed. Most of them have either silted up or have been drained.

6. Geology.

On studying the origin of the existing geography of the East Riding we find that the region consists exclusively of sedimentary rocks. It is therefore clear that, before it could exist, other regions must have been gradually worn down to supply the materials, most of which were deposited in the sea, though in some cases in lakes and estuaries. The strata so formed were then, from time to time, elevated above sea-level, tilted in some cases out of their original horizontal position, and then themselves exposed for long ages to subaerial denudation, by which the valleys were carved out and the inter-stream uplands were left as hills, in some cases because they consisted of more resistant strata, in others only because they were less vigorously attacked than the valley bottoms. In the Great Ice Age vast quantities of materials, transported by ice, were deposited on the lowlands or in the shallow neighbouring sea and so raised the level of its floor as to convert it into dry land.

The annexed table of the geological formations represented in the East Riding is based on the Memoirs of the Geological Survey. The areas were measured by planimeter, on the one-inch-to-the-mile Geological Survey maps, by Mr H. Dewey.

Carboniferous, Permian, and Trias.

The oldest rocks which are known to occur in the East Riding are of Carboniferous age, for a boring south of Selby reached the Coal Measures at a depth of 1284 feet, so that those measures no doubt extend into the East

Geological Formations represented in the East Riding

			Formation	Thickness in Feet		Area in square miles
QUATERNARY OR POST-TERTIARY		RECENT AND POST-GLACIAL	Blown Sand			
			Recent Alluvium, Modern Warp, Peat		mainly warp clay	267
			River Terraces			
			Ancient Warp and Lacustrine Clay			
			Sand and Gravel			153
			Gravel of Dry Chalk Valleys			5
			Chalky Gravel (capping hills) and Gravel Terraces			
		PLEISTOCENE OR GLACIAL	Glacial Sands and Gravels			33
			Boulder Clay			356
SECONDARY OR MESOZOIC	CRETACEOUS	UPPER CRETACEOUS	Upper Chalk	1200+		
			Middle Chalk	221		294
			Lower Chalk	60 to 123		
			Red Chalk (a red marl)	2 to 30		
		LOWER CRETACEOUS	Speeton Clay*	330	Absent in the south	
	JURASSIC	UPPER OOLITE	Portlandian ? part of zone of *Belemnites lateralis* at Speeton 35 ft.	Under 100 to 500+		
			Kimmeridge Clay*			
		MIDDLE OOLITE — CORALLIAN	Upper Calcareous Grit and North Grimston Cement Stone Beds (argillaceous limestone)	45 to 75	Absent in the south. Area exclusive of Oxford Clay	8
			Upper Limestone and Coral Rag	100		
			Middle Calcareous Grit (Filey Brig Grit)	16		
			Greystone or Passage Beds (siliceous subolitic limestones and grits)	35		
			Lower Calcareous Grit	30 to 60		
		LOWER OOLITE	Oxford Clay* (150 ft. near Filey)	20 to 70		
			Kellaways Rock (soft sandstone	10 to 40		
			Upper Estuarine Series (sandy beds)	10 to 70	Absent in the south	7
			Scarborough or Grey Limestone Series	30		
			Middle Estuarine Series (sandy rock with shale or clay)	20 to 30		
			Millepore Series or Cave Oolite (oolitic limestone)	10 to 30		
			Lower Estuarine Series (sandstones and shales but including Hydraulic Limestone)	20 to 100		
			The Dogger (sandstone or ironstone)	12		
		LIAS	Upper Lias (shales)	90		
			Middle Lias (ironstone, sandstone and limestone)	3 to 9		24
			Lower Lias (alternating limestones and shales)	100 to 150		
	TRIAS		Rhætic Beds (shales with sandy beds)	15?		
			Keuper Marl	634 + about 200 (boring)		14
			Keuper Sandstone	68		
			Bunter Sandstone	1425		
PRIMARY OR PALÆOZOIC	PERMIAN		Upper Permian Marl	212		
			Upper Permian Limestone (magnesian)	141		
			Middle Permian Marl	29		
			Lower Permian Limestone (magnesian)	over 589	(to bottom of bore hole)	

* The Oxford, Kimmeridge and Speeton Clays have a combined area of 13 square miles. The geographical distribution of the different formations is shown by the geological map at the end of the book but it only shows the solid geology. For the Glacial Drift see the sketch-map, page 30.

Riding. The oldest rocks actually observed in the riding
are of Permian age and are known only from a boring
about 2½ miles S.W. of Market Weighton. After pene-
trating 2 feet of superficial deposits, the bore entered the
Keuper Marl about 200 feet below its summit and, with
the exception of the Rhætic, all the Triassic and Permian strata shown in the table, with the thicknesses
there given were met with. The superficial deposits of the
Vale of York are underlain by Triassic sandstone to west
of a line joining Wilberfoss and Howden, and by Keuper
Marl to east of that line, and the red and grey Keuper
Marl is well exposed near the Derwent, 1½ miles N. and
S. of Scrayingham. The Permian and Triassic strata were
probably deposited, under desert conditions, in a closed
inland sea like the Caspian, or partly as subaerial river
deltas. The Rhætic and Liassic beds are marine.

JURASSIC.

Rhætic and Jurassic strata underlie the whole of the
East Riding north and east of the Vale of York, but
are only exposed near Filey*; at the foot of the Wolds
S.W. of Knapton*; in the Howardian Hills, and at the
foot of the Wolds from Acklam to within a mile of the
Humber. In the Lias, ammonites (belonging to the same
sub-class as the existing Pearly Nautilus) and belemnites
(allied to existing cuttle-fish) appear for the first time in
England and persist until the close of the Cretaceous. In
the Lias also remains of marine reptiles have been found—
Ichthyosaurus at North Cliff and Plesiosaurus at Cliff south
of Market Weighton. Mr C. Fox Strangways suggested
that land may have existed as a promontory separating the

* Jurassic only.

[north east] Yorkshire basin from the Liassic sea of South Yorkshire and the locality of this supposed promontory coincides with that of the arch of the Market Weighton anticline.

With the incoming of Oolitic times we meet with a great change in physical conditions. The Dogger is still marine but above it the Lower Oolites of Yorkshire, unlike those of the South of England, consist of estuarine deposits, separated in places into three successive series by intervening marine strata. The conditions were very similar to those which prevailed in Lower Coal Measure times. Low-lying marshy areas of alluvium supported a luxuriant vegetation.

With the oncoming of the Middle Oolites in the Kellaways Rock (rich in fossils east of South Cave station) we return to marine conditions. The Corallian grits and limestones were only deposited to the north of the Market Weighton anticlinal arch. The Upper Oolites are also marine. The Kimmeridge Clay underlies the superficial deposits of the whole of the Vale of Pickering.

At the close of Jurassic times those forces which tend by folding of the strata to produce mountains came into play and the Jurassic rocks were thrown into folds, trending east and west, thereby giving rise to the synclinal basin of the Vale of Pickering. There is no doubt that before the deposit of the Red Chalk, the Jurassic rocks must have suffered great denudation and that the Cretaceous rocks overlie their upturned and truncated edges in most places unconformably.

CRETACEOUS.

The Speeton Clay is of special interest as it affords a more complete succession of marine deposits of Lower Cretaceous age than occurs in any other part of England except Lincolnshire. It is best exposed in Speeton Cliffs, Filey Bay, where it is succeeded by the Red Chalk (containing *Belemnites minimus*) which is of the same age as the Gault of southern England. Above it is the Chalk proper, a white limestone, though some beds of the Lower Chalk in the Speeton Cliffs are pink. The Chalk is marine and, like most limestones, is largely of organic origin. The Lower Chalk is usually free from flints, but beds of flint some yards in extent occur in it at Speeton Cliffs. In the Middle Chalk flints appear 8 feet above its base and continue throughout. In the Upper Chalk flints occur in the lower 345 feet.

The Wolds are composed of Chalk, of which there are several outliers—that is isolated portions—separated from the main mass or escarpment by denudation. The finest exposure of the Chalk is in the cliffs of Flamborough Head. (See illustration, p. 45.)

POST-CRETACEOUS ELEVATION, DENUDATION, AND RIVER DEVELOPMENT.

Pebbles of quartz have been found on the Wolds, which may indicate the former existence of overlying Tertiary strata, but, with this possible exception, the Eocene, Oligocene, Miocene, and Pliocene systems are unrepresented by deposits in the East Riding.

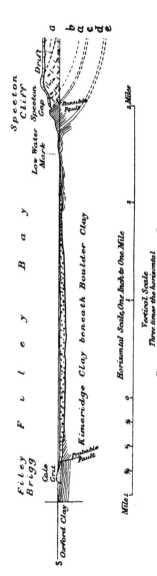

Section from Filey Brigg to Speeton Cliff (*C. Fox Strangways*)

a. Chalk without Flints.
b. Chalk with Flints.
c. Red Chalk.
d. Speeton Clay.
e. Passage Beds.
S. Sea level.

Vertical Scale
Three times the horizontal

Feet 1000 500 0 1000 Feet

Horizontal Scale, One Inch to One Mile

Mile ¾ ½ ¼ 0 ¼ ½ ¾ 1 2 3 Miles

The Mesozoic rocks of the East Riding originally extended far to the west of their present escarpments. The

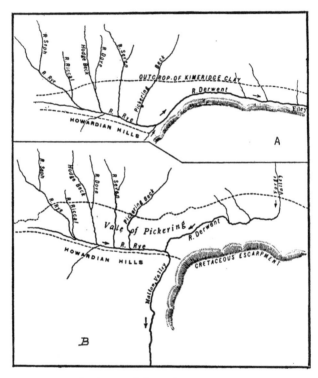

Drainage of the Vale of Pickering

whole of the East and much, if not all, of the North and West Ridings were covered by a continuous sheet of Upper Chalk. In early Eocene times this area, together with a great part of the British Isles, was elevated above sea level and became dry land, with a slight south-easterly inclination, which would be increased by the final elevation of the Pennine Chain in late Eocene or early Oligocene times. In late Oligocene or early Miocene times an uplift along the east and west axis of the Moorland Range (between the Tabular and Cleveland Hills) gave rise to a new watershed and imparted a southerly dip to the Cretaceous rocks on its southern flanks. A new set of secondary consequent streams (that is, streams running in the direction of the dip or inclination of the strata), *e.g.* the Hodge Beck, Dove, and Seven, was developed. Meanwhile the covering of Chalk was being denuded, and in the Vale of Pickering a subsequent river (following the strike of the strata) would arise, following the general direction of the present Hertford River but flowing eastwards. (Stage A of diagram opposite.)

It was during the vast period of time between the Lower Eocene and the Pleistocene Glacial epoch that the enormous erosion and denudation took place, which not only removed the Chalk from the Vale of York and the Vale of Pickering, but excavated the broad Vale of York in the soft Triassic rocks, and the Vale of Pickering in the soft Kimmeridge Clay and cut back the escarpments of Jurassic and Cretaceous rocks to nearly their present positions.

That some time before the Glacial epoch the land stood at very nearly its present level above the sea is shown by the deposits banked against the buried sea cliff of Chalk at Sewerby, near Bridlington Quay. The section there shows *inter alia* in descending order:—

1. Basement Boulder Clay (Glacial).
D. Chalk-rubble.
C. Blown Sand.   Each with mammalian
B. Land-wash of marly clay   bones and land shells.
    and fallen chalk.

A. Old Sea Beach of rounded pebbles, with its upper surface not much above the present level of the highest tides, and containing marine shells and bones of *Elephas antiquus*, *Rhinoceros leptorhinus*, *Hippopotamus amphibius*, hyæna, bison, etc.

The Chalk-rubble has been proved by borings to descend to at least 23 feet below high water mark, indicating that between the period represented by the beach and that of the Basement Boulder Clay, the land must have stood at a higher level than it now does. As already mentioned (Chapter 4, p. 7) this buried cliff is continued through Bridlington, Driffield, Beverley, and Cottingham to Hessle, and it shows that when it was formed Holderness did not exist, but that the sea in a great bay occupied its place.

Pleistocene or Glacial.

The Glacial deposits of the East Riding consist of Boulder Clay and Sands and Gravels. Boulder Clay is a clay containing ice-transported boulders of rock and is believed to be either the bottom moraine of glaciers or the residuum left behind on the melting of a sheet of glacier-ice charged with clay and stones. In the cliffs of Holderness Mr Clement Reid recognised four successive boulder clays separated by sand and gravel—(1) Basement Boulder Clay with ice-transported masses of sand containing marine

Scale: Horizontal & Vertical.

5 10 20 30 40 FEET.
10 20 30 40 50

100 Yards.
300 Feet.

S.W.

N.E.

Boulder of Grit
in Chalk-rubble.

Bridlington WATER MARK.
Cliff of
Chalk.

Index.

Glacial Drifts.

4 — The Sewerby gravels (late glacial.)

3 — "Purple" Boulder-clay splitting into two bands, South-westward, with stratified material intervening.

2 — Stratified clay, loam, sand & gravel, nearly thinning out North-Eastward.

1 — "Basement" Boulder-clay including contorted patches of loam, chalk-rubble, etc. towards base.

Infra-Glacial Beds.

D — Chalk-rubble. (see text)

C — Blown Sand, with bones, etc.,(a land-shell found in a loamy band near top.)

B — Land-wash banked against the buried cliff, with bones & numerous land-shells.

A — Old Sea-beach, with bones & marine shells.

Section of Buried Cliff at Sewerby near Bridlington Quay (G. W. Lamplugh)

shells, mostly arctic. (2) Lower Purple Boulder Clay. (3) Upper Purple Boulder Clay. (4) Hessle Clay (of the sea-coast section). The distribution of the Glacial deposits is shown on the sketch-map, p. 30.

The sequence of events in the Glacial epoch is believed to have been as follows :—an influx of glacier-ice from the Clyde forced the glacier moving down the Vale of Eden to reverse its direction of flow and cross the watershed of the Pennine Chain by the Stainmoor Pass and flow down the Valley of the Tees, transporting boulders of Shap granite and other rocks. This Teesdale glacier must have discharged into the North Sea, flowing south-eastward along the coast, as Shap boulders are numerous at Robin Hood's Bay.

The second phase of the glaciation of East Yorkshire was the complete diversion of the Teesdale ice into the Vale of York, where it has left two great concentric crescent-shaped terminal moraines; one followed by the road from York towards Stamford Bridge, the other curving by Wilberfoss, Wheldrake and Escrick. The cause of this diversion was the pressure exerted upon it by a great Scandinavian glacier, which crossed the North Sea, obstructed the access of the Teesdale ice to the coast, and transported boulders of Scandinavian rocks to Holderness. The third phase of the glaciation was the arrival off the East Riding coast of ice from the valley of the Tweed and the Cheviot Hills.

We may here mention the remarkable history of the River Derwent. It is a compound river formed of portions which had originally no connection one with the other. The upper Derwent originally flowed to the sea by the

East Syme valley (North Riding); this was blocked by
glacier-ice, a lake called Harwood Dale Lake was formed
which overflowed and excavated Langdale Gorge; similarly the valley of Sea Cut (Scalby Beck) was ice-blocked,
Harkness Lake was formed, overflowed, and excavated
Forge Valley; lastly the valley of the pre-glacial Hertford
River (flowing into the sea at Filey) was blocked by
glacier-ice and the great Lake Pickering was formed
which overflowed near Malton and excavated the beautiful
gorge of the Derwent below Malton. When this had
been cut deep enough the lake was drained and the Hertford with flow reversed joined the Derwent from Forge
Valley (see Diagram, p. 24, stage B) which escaped by the
Malton gorge into the Vale of York and the extra-
morainic glacial Lake Humber on the west side of the
Chalk escarpment which was held up by glacier barriers
at each of its exits.

When the glaciers which had invaded the Yorkshire
coast finally melted away, they left Holderness a land of
moraine meres, mounds of boulder clay and of so-called
Interglacial Marine gravels (containing marine shells) as
at Kilham, Brandesburton Barf, Kelsey Hill (near Burstwick), etc.

POST-GLACIAL AND RECENT.

Of the post-glacial deposits of Holderness perhaps the
lacustrine deposits of the ancient meres, which have been
exposed in depressions in the cliffs and on the foreshore by
coast erosion, are the most interesting. They consist of peat
with plant remains, clay and marl with fresh-water shells

and gravel. The best example is at Hornsea, where the
deposits are continuous with those of the existing mere.
There, and in the lake deposit at Skipsea, remains of the
Irish elk, *Cervus megaceros*, have been found. Other
localities are on both sides of Bridlington Quay, where

Geological Map of the East Riding showing Glacial Drift

leaves of the Arctic birch, *Betula nana*, have been found,
as also at Holmpton (though there is not much to be seen
there or at Out Newton, Atwick, and Barmston). At
Sand-le-Mere (1 mile S.E. of Tunstall), on the foreshore,
not only lacustrine but estuarine deposits with shells, *e.g.*

Hydrobia, have been found. Withernsea, with remains of red deer and reindeer, and Owthorne are other localities.

In the excavations for the Albert Docks, Alexandra Docks, and Alexandra Dock Extension at Hull a buried forest, consisting of a bed of peat, with stumps of oak trees and remains of bird-cherry, birch, and hazel was found, overlying boulder clay at a depth (in the third locality) of 13 feet below ordnance datum or 25 feet below high-water mark of ordinary spring tides. This shows that, after glacial times, the land stood at a higher level than at present, so that the Humber estuary is a drowned river valley. There is a submerged forest on the Humber shore at Melton, near North Ferriby. Much of the country on the banks of the Humber consists of estuarine warp, the tidal alluvium deposited by the Humber. In the Vale of York the superficial deposits are from 25 to 90 feet thick. A very large area is there covered by superficial clean yellow and white sands, from a few inches to 15 feet thick. Beneath these sands there usually occurs a thick deposit of laminated clay (lacustrine clay) or of ancient estuarine warp. Along the sides of the Ouse and Derwent there is a narrow strip of modern alluvium.

In the Vale of Pickering, at the foot of the Chalk escarpment, there are deposits 30 to 50 feet thick of sands, including fragments of chalk and flints. Nearer the Derwent are deposits of warp and lacustrine clay. All these deposits in the Vale may be partly pre-glacial, partly glacial and post-glacial. Between Ayton and Sherburn, a deposit of modern alluvium flanks the old course of the Derwent, forming a slight ridge, at a higher level than the surrounding country. This must have been deposited during floods.

7. Natural History.

Islands may be divided into two chief classes—oceanic and continental. Oceanic islands are chiefly of volcanic or coral-reef origin and have never formed part of any continent or large mass of land. Continental islands are detached fragments of continents from which they have been separated by submergence or erosion. Great Britain and Ireland are continental islands of geologically recent origin ; that is, their separation from the Continent is recent. That the British Isles have formed part of continental Europe is shown not only by the fact that remains of the Pliocene and Pleistocene species of mammals of the British Isles occur as fossils on the Continent, but also by the fact that the existing animals and plants of these islands closely agree with those of the Continent. On the other hand, if the separation had existed for a long period many peculiar British forms would certainly have been evolved, which is not the case. The separation is believed to have taken place at the Straits of Dover in late Pleistocene times.

Although we know that the same temperate species of plants that live in Great Britain now were here in pre-glacial times, it is uncertain to what extent the pre-glacial plants of Britain survived the Pleistocene glaciation in this country. It has been recently shown that many moorland plants of Yorkshire occur in arctic regions, so that it is quite possible that they survived the Pleistocene glaciation. Owing to the limited area, the restricted range of climatic and other conditions in the British Isles, and no doubt also to the short time which elapsed after the glacial epoch before separation from the mainland occurred, Britain is less rich in species of both animals and plants than the European continent. The vast majority of the British animals and plants have been derived from the Continent, and the few forms peculiar to Britain are probably due to changes brought about by isolation.

The distribution of vegetation is chiefly determined by soil conditions and climatic conditions. The former depend primarily on the geology, but in cultivated areas are modified by human agency; the climatic conditions depend mainly on altitude and on the temperature and rainfall, which, in their turn, are closely connected with height.

One of the striking features of the East Riding is that only a very small part of its area is in its original, natural, uncultivated, and undrained state. In 1914 there were 308,264 acres of mountain and heath land used for grazing in the North Riding as compared with 232,187 acres in the West Riding and only 2063 acres in the East Riding, but in 1922 the East Riding area was 8863 acres. This subject is referred to below under "Derwentland."

According to Mr J. F. Robinson, the chief authority on the botany of the East Riding, the riding may be divided into three botanical areas or districts, viz.—I, Holderness in the wide sense ; II, The Wolds ; III, "Derwentland," *i.e.* the Plain of York and the Vale of Pickering.

I. In Holderness Spurn Head, five furlongs wide at Kilnsea, has a length of 4 miles and is generally under 300 yards wide. It consists of sand-dunes covered by a dense growth of marram and other grasses with Sea Buckthorn (*Hippophaë rhamnoides*) and Sea Holly (*Eryngium maritimum*), etc.

On the north or left bank of the Humber four successive botanical zones or strips may be distinguished—(1) The Mud Flats, with no flowering plants but with green algæ, which hold the mud together. (2) The Salt Marsh, submerged only at spring tides, with abundant flowering plants such as the Sea Purslane (*Atriplex portulacoides*), Sea Lavender (*Statice Limonium*), etc. (3) The Salt Meadow, at an average height of 3 or 4 feet above the salt marsh, with Thrift (*Armeria maritima*), the Mud-Rush (*Juncus Gerardi*), etc. (4) The landward edges of the salt meadow and both faces of the artificial embankment, with Sea Clover (*Trifolium maritimum*) and abundant Sea Wormwood (*Artemisia maritima*), etc.

The plants of the interior of Holderness may be divided into terrestrial and aquatic. The terrestrial plants are chiefly clay-loving (pelophilous) when they grow on Boulder Clay, or dry-loving (xerophilous) when they grow on gravel mounds, such as Brandesburton Barf. As a locality for aquatic plants Pulfin Bog (marsh), 40 to 50 acres, a western extension of Eske Carrs, N.N.E. of Beverley, may be mentioned. It has a profuse growth of reeds (Phragmites), bulrushes, iris, etc. Hornsea Mere is another locality for aquatic plants.

II. The Wolds, being composed of chalk, are naturally characterised by xerophilous or by calcophilous (lime loving) plants such as the common Rock-rose (*Helianthemum Chamæcistus*), the Dropwort (*Spiræa Filipendula*), a St John's Wort (*Hypericum montanum*), the Beech, etc.

III. "Derwentland" is chiefly a tract of alluvial and blown sand though aquatic plants occur in the rivers and streams. Mr J. F. Robinson informs me that the principal uncultivated areas in the Vale of York are Skipwith Common, about 1000 acres, and the adjoining Riccall Common; Barmby Moor, 500—600 acres, with *Erica cinerea* and *E. Tetralix*; Allerthorpe Common, 600 to 700 acres with Birch; Hotham Carrs 300 acres; Holme on Spalding Moor, not 120 acres; and Market Weighton Great Sand Field, odd patches, the latter with Sundews (*Drosera*), and the last two localities with Dwarf creeping Willow (*Salix repens*). All the preceding "Derwentland" localities are characterised by the Common Ling (*Calluna Erica*). Of South Cliff Common odd patches only and of Bubwith Ings very little is really wild. The area of woodland in the riding is small—18,473 acres in 1913.

Space will not permit more than a sketch of the vertebrate fauna of the East Riding. Beginning with the mammals we have the long-eared bat, noctule, pipistrelle, Daubenton's bat, and the whiskered bat. The hedgehog, mole and common shrew are common, the water shrew and lesser shrew are rare. The fox increased greatly in numbers during the great war, the stoat and weasel are common. The badger holds its own on the Wolds, in spite of much persecution. The otter occurs at Turnhead Reach on the Ouse, Menthorpe Ferry on the Derwent, at Driffield, Scampston, etc. In Neolithic times the beaver occurred in the riding. The common seal sometimes occurs on the coast, and two were captured at Naburn in the Ouse in 1891. The squirrel is not common, except in the plantations on the Wolds. The brown rat, water vole, long-tailed field vole, short-tailed field vole and house mouse are common. The black rat is rare, but occurs in drains and basements in

Hull. The red bank vole is widely distributed but hardly common. Rabbits are very common and hares are common, particularly on the Wolds. Cetaceans are chiefly occasional visitors, but porpoises are common and, like the grampus, follow the salmon up the River Ouse.

Among the rare cetaceans are four species of rorqual, the sperm whale cast ashore at Tunstall in 1825, the

Great Crested Grebe

common beaked whale (*Hyperoödon rostratus*), the Beluga or white whale, pilot whale (*Globicephalus melas*), white-beaked dolphin and bottled-nosed dolphin. By an Inquisition taken 6th Edward I (1277–1278) Gilbert de Gaunt and Richard de Malebisse are declared to have the right to take a whale whenever caught in their port of Fiweley (Filey), the head and tail only being reserved for the king.

The number of species of birds which have been recorded

as occurring in Yorkshire is 325, of which 123 are
annual breeders, and 202 migrants or occasional visitors.
Most probably almost all of these occur in the East
Riding. Of these, 229 species (including 53 resident
species) occur at Flamborough and of those recorded at
Flamborough 117 species also occur at or near Hornsea
Mere.

Nest of Little Grebe

At Hornsea Mere there is a heronry with between
30 and 40 nests; another exists at Stillingfleet. At
Hornsea Mere too the great crested grebe breeds, its
nest among the reeds being remarkable for its "sea-
worthiness" at fluctuating water-levels. The little
grebe or dabchick, the reed-warbler, the kingfisher,

the tufted duck, the mallard and the water-rail nest
there. The brown owl is common there and a pair or
two of barn owls usually nest near the mere. In the
water holes of Skipwith Common the black-headed
gull has increased until there are at present (1922) about
400 pairs. In the same locality "the mallard, common
teal, shoveller, common snipe and redshank also breed

Ringed Plover

regularly." The Spurn peninsula is the breeding ground
of some interesting birds, though the species are not
numerous. The ringed plover and little tern are the chief
species, but in addition yellowhammers, linnets, reed bunt-
ings, etc. occur.

At Flamborough Head the sea birds are of most interest
and their eggs are collected every year by adventurous

Egg Collecting, Bempton Cliffs

native climbers who descend the vertical cliffs between Selwicks Bay and Speeton Cliffs by means of ropes, and collect an estimated number of 80,000 eggs annually, part for food, part for clarifying wine and dressing patent leather, and some for ornithologists. The chief species breeding there are the guillemot, razorbill, and puffin, but in addition the kittiwake, herring gull, rock dove and stock dove, jackdaw, starling, rock pipit, house martin, carrion crow, kestrel and occasionally peregrine falcons are resident, and many other birds are visitors. The razorbill is far less numerous than the guillemot at Flamborough and prefers to lay its egg in holes or crevices instead of on the open ledges.

Of reptiles the common lizard (*Lacerta vivipara*) occurs at Filey, Pocklington, and Spurn; the blindworm and viper occur on Skipwith Common; the grass snake, as well as the viper, is recorded from Pocklington and is spreading all over Holderness.

8. Peregrination of the Coast.

The length of the coast-line of the East Riding is sixty miles and the area of its foreshore is 3750 acres, therefore the average width of foreshore is 172 yards. The chief features of the coast are Carr Naze with Filey Brig, Filey Bay, Flamborough Head, Bridlington Bay, and Spurn Head.

Beginning on the north, at the Wyke, the vertical cliffs, called the North Cliff, form an almost straight line running E.S.E. in the direction of strike of the beds, and continue uninterruptedly into Carr Naze. On the north side of

Carr Naze they have been worn by the sea into several rock-bound almost circular coves, called " doodles."

Filey Brig is a continuation of Carr Naze, from which the glacial drift has been removed by marine denudation. It extends for nearly half a mile beyond Carr Naze, as a natural breakwater, exposed only between tide marks, and forms a dangerous reef, protected by a bell buoy. In

Doodle or Cove on North side of Carr Naze

stormy weather the waves break over it and the spray shoots upward in great jets. The force of the waves detaches huge blocks of the Corallian grit, which are piled up in confusion on both sides of the Brig. The writer measured one 8 feet square by 2 feet 6 inches in thickness, which would weigh about eleven and a half tons. Carr Naze and Filey Brig owe their prominence

to the great hardness of the Corallian grit, compared with
the Oxford Clay below it and the Kimmeridge Clay and
Glacial Drift of Filey Bay above it.

Filey Bay, on the other hand, owes its existence to the
easily eroded boulder clay. It extends in a gentle curve
of 6 miles to Dulcey Dock and a further mile and a

Filey Brig. Grit in the foreground, the Glacial Drift
of Carr Naze in the background

half before the tide-bound cliffs of Bempton bar access to
their foot. On the south side of Carr Naze boulder clay,
carved by rain and weather into ridges and pinnacles,
forms the upper part, and gently sloping beds of Corallian
grit the base of the cliff, and the same beds of grit form
the beach, but, as we proceed towards Filey, these give

place to a broad beach of sand, and the cliffs consist entirely of Glacial boulder clay, sands, and gravels. Filey is beautifully situated on the top and seaward slope of boulder clay cliffs just over 100 feet in height. It is a comparatively quiet watering place without the wealth and fashion of Scarborough or the day trippers of Bridlington. It has no

Weathered Boulder clay overlying Corallian rocks, Carr Naze

pier or harbour, but there is a life-boat station and a promenade and sea-wall 700 yards long, built in 1892.

Flamborough Head, the seaward end of the Wolds, forms a blunted triangle, each side of which measures 5 to 6½ miles. Like Filey Brig, its existence as a promontory is due to the superior hardness of its Chalk cliffs, compared with the Glacial drift of Filey Bay and Bridlington Bay. Its splendid cliff scenery contrasts with that of the gently undulating surface of the Wolds. The upper part of its cliffs is composed of Glacial Drift. At the south-eastern side of Filey Bay the Speeton Clay gives rise to numerous landslips and screes, which in Speeton cliffs (440 feet, far loftier than the east end of Flamborough Head) form an undercliff of shattered chalk mixed with drift-clay. In Buckton and Bempton cliffs the vertical walls of white chalk form magnificent precipices, accessible only to sea-birds and the hardy climbers who, with the aid of ropes, collect their eggs. As Dr A. W. Rowe remarks, "the chief glory of the Yorkshire coast lies in its bays. Thornwick, North Sea Landing, and Selwicks display a beauty so rare that they compel the admiration of the most careless; and that beauty is enhanced by the fact that the presence of these bays is so unexpected." The sides of all the bays of Flamborough Head consist of vertical walls of chalk, capped by drift. A little over a mile S.E. by E. of the northern end of Danes' Dyke is Sanwick, an obtuse-angled recess rather than a bay. Immediately east of it is Chatterthrow, a narrow inlet with two natural arches, communicating with the next narrow inlet, Little Thorn-wick, which unlike the preceding ones is accessible, and has a natural arch on its east side. Great Thornwick is a much larger bay than those previously mentioned. It is very accessible and has the Church Cave and Smuggler's Cave on its east side and a fine natural arch on the west. A quarter of a mile east of Great Thornwick is North Sea Landing, a natural harbour for small boats and a life-boat station. It has a double natural arch (Bacon Flitch Hole) on the west side and Robin Lythe's Hole, the chief cave of Flamborough, on its east side. Three furlongs from

North Sea Landing is South Breil in which are two isolated sea-stacks, one of which is also a natural arch; they are called the King Rock and the Queen Rock. Nearly a mile S.E. of South Breil is Selwicks Bay, with several natural arches and two chalk stacks, called Adam and Eve. On the south side of the bay there is a deep crater-like pit in the Glacial Drift of the cliff, the upper opening

East sides of Chatterthrow, Little and Great Thornwick Bays, Flamborough Head. (*Chalk capped by Drift*)

of a "blow-hole" called Pigeon Hole, formed by the collapse of part of the roof of a sea cave in the chalk, which has penetrated a pre-glacial Drift-filled valley. Only 380 yards S.E. by S. of this is another similar "blow-hole" (in the same pre-glacial valley) in High Stacks, the eastern-most point of Flamborough Head (under 125 feet high), and close by is a fine natural arch.

At the southern angle of Selwicks Bay is Flamborough Lighthouse, its light visible 21 nautical miles. Near it is a coastguard station and an ancient octagonal look-out tower.

The south side of Flamborough Head is much less interesting than the north side. Its height is only from 75 to 125 feet (except at Beacon Hill 180 feet) and it exhibits only two indentations worthy of mention, both lined with drift—South Sea Landing with coastguard and life-boat station and Danes' Dyke ravine. The southern beach is much more accessible than most of the northern shore. Bridlington Bay is well defined on the north, but to the south-east has no definite boundary.

Opposite Sewerby the buried sea-cliff described in Chapter 6, p. 25, occurs, and the coast to the south of that point for 37 miles, as far as Kilnsea, consists of Glacial Drift, the feeble resistance of which to erosion accounts for Bridlington Bay. About a mile and a half south-west of Sewerby is the large and popular watering-place of Bridlington Quay. It has a sea-wall 2845 yards long, a life-boat station, and an ancient harbour of about 12 acres in extent, dry at low spring tide, enclosed between two stone piers, that on the north 235 and that on the south 500 yards long. The coast south of Bridlington to Kilnsea consists of cliffs of variable height, composed of reddish or chocolate-coloured, and sometimes dark green boulder clay, with intercalated bands of sand and gravel. The foreshore consists of sand and shingle, and occasionally boulder clay. The traces in the cliffs of ancient meres have been mentioned in Chapter 6, p. 29. Twelve miles south-eastward of Bridlington is Hornsea, a small watering place with a concrete sea-wall and promenade, 310 yards in length, and a life-boat station. Hornsea had an iron pier until about 1887, when it was washed away. The chief attraction of Hornsea is its Mere, described in Chapter 5, p. 17.

Fourteen and a half miles south-eastward of Hornsea is Withernsea, a rapidly growing watering place, very popular with the inhabitants of Hull, on account of its nearness and bracing climate. It is protected by a concrete sea-wall, 1540 paces long, with a promenade on top and has a lighthouse. It had an iron pier 1196 feet long, opened in 1877. In October 1880 the pier was swept through by a drifting vessel and in 1893 was irreparably damaged.

Two and a half miles S.E. of Withernsea, the Runnell reaches the coast. The cliffs here are only 6 to 8 feet high from the beach and are capped by a freshwater (Holmpton mere) deposit with peaty layer. On the other hand, near Out Newton, at 5 miles S.E. of Withernsea is Dimlington High Land (Beacon) with a cliff of boulder clay and gravels over 100 feet in height. At Easington there is a life-boat station. Four miles S.E. of Dimlington High Land is Kilnsea, only 17 feet above ordnance datum, where the low-lying peninsula of Spurn may be taken to begin. It is four miles long, and at Kilnsea five furlongs broad, but is usually under 300 yards wide. It consists of the detritus, eroded by the sea from the long stretch of boulder clay and gravel cliffs north-west of it, and swept down the coast by the southward drift of the shore-deposits. The surface, above tide marks, consists of sand-dunes, as described in Chapter 7, p. 33, which overlie boulder clay. The peninsula does not consist exclusively of sand. There is shingle exposed on both sides. At the southern end is a life-boat station and a lighthouse which in 1895 replaced that of Smeaton. The coast of the Humber has been described in Chapter 4, p. 11, and Chapter 7, p. 34.

9. Coastal Gains and Losses.

The coast of the East Riding, except where protected by sea-defences, is undergoing marine erosion, though at Spurn deposition occurs.

The rate of erosion at Filey Brig is probably slow. In Filey Bay there is an average loss of 2 feet per annum. According to Mr E. R. Matthews, Flamborough Head is receding at 1 foot 6 inches per annum. On the other hand, the Holderness coast is being more rapidly eroded than any other part of Britain. Mr C. Thompson has measured, on the ground to the edge of the cliff, in 1922, sixty lines, shown on the six-inch Ordnance Survey Map of 1852 and, by comparison with that map, ascertained the gain or loss. At Auburn, 2¾ miles S.S.W. of Bridlington, he finds that the land gained 40 feet prior to 1852 and 50 feet since. The following measurements are the average loss per annum, unless otherwise mentioned.

At Barmston Drain, 2½ miles south of Auburn, the loss is 2 feet; at Ulrome nearly 5 feet; at Skipsea 5·92 feet; south of Skipsea to Hornsea sea-defences 3·14 feet to 4·37 feet; Hornsea Burton to Mappleton Gap 3·79 feet to 4·7 feet. At Aldbrough Beer House the loss is 4 feet per annum, although the cliff proper has not been eroded during perhaps the last ten years. At Sand-le-Mere the loss is 4 feet; south of Withernsea 4 feet 6 inches; at

Holmpton over 5 feet. Dimlington farm has suffered very severely, having lost at least 9 feet per year.

Between Easington and Kilnsea the erosion has been very rapid, *e.g.* (proceeding southwards):—

	Total in 70 years	Annual
Kilnsea Beacon	563·5 ft.	8·5 ft.
Two fields S. of Beacon	1300	18·57
Maximum loss at top of 6 inch sheet, No. 269	1450	20·71
From edge of cliff to line of dated stone in Blue Bell Inn, Kilnsea	673·5	9·62

The land lost, however, in this tract, was mostly low-lying marshy fields, and the greater part of the loss is due to flooding during periods of very high tides. Mr T. Sheppard remarks: "In 1905–6 the lowlands from Kilnsea Warren nearly to Easington Lane end on the north, and beyond Skeffling on the west, were flooded in consequence of the sea breaking through the artificial banks along the coast."

Mr C. Thompson finds for the Holderness coast generally (omitting Kilnsea) an average loss of 4·8 feet per annum. This is less than the 7 feet per annum usually given as the average loss of the coast, but this latter figure includes Kilnsea.

According to Mr W. Shelford, Spurn Head, between 1676 and 1851, grew southward 2530 yards or 14½ yards a year and Mr A. E. Butterfield finds that, between 1851 and 1888, the high-water line, at Spurn, has advanced 200 yards southward or nearly 17 feet per annum. He also finds that Spurn has been increasing in width and extending westwards. Between 1851 and 1888 the

westerly movement on the North Sea side has been
99 yards, or 8 feet a year, whereas in the same 37 years,
the westerly movement on the Humber side has been
210 yards or 17 feet a year; a gain in width of the land
of 111 yards.

As the result of marine and Humber erosion many
places have disappeared. These have been described as
"lost towns" but, in most cases, they were mere villages
or even hamlets, although the area of land lost is large.
It is to the late Mr J. R. Boyle that we are chiefly
indebted for information as to the losses on the
Humber, and to Mr T. Sheppard as to those on the
sea-coast.

From an inquisition, held at Hedon in 1401, we learn
that the abbey of Meaux (pronounced *muce*, $3\frac{1}{2}$ miles E.
of Beverley) had lost, by the overflowings and consump-
tions of the sea and of the water of the Humber, in
Tharlesthorpe 573 acres and 4 messuages; in Saltagh
$282\frac{1}{2}$ acres: in addition, land in Wythfflete (Orwithfleet),
Frysmersk, Drypule (Drypool) and Otringham (all the
preceding places being on the Humber). On the sea-coast
the abbey lost lands in Dymelton (Dimlington), Grymeston
(Grimston), Coldon (Colden or Cowden), Hornseburton
(Hornsea Burton), Hyth (E. of Skipsea), Skypse (Skipsea),
Ulram (Ulrome), Cleton (S.E. of Skipsea), Wythornese
(Withernsea), and Hertburne (Hartburn on the Earl's
Dyke, near Bridlington).

Drypool of course is in the city of Hull and Ottringham
is well known, but the exact position of the other localities
on the Humber is uncertain, but according to Mr T.
Sheppard, Saltagh and Tharlesthorpe were between Paull

and Keyingham; East Somerte* S. of Keyingham; Orwithfleet S. and Frysmersk S.E. of Ottringham; Penisthorpe* W. of Welwick. The alien Priory of Birstall (or
Burstall) was situated three quarters of a mile south of
Skeffling. Its ruins still existed in 1721 but have been
swept away by the Humber.

The most interesting places which have disappeared
are Ravenser and Ravenser Odd; the latter place, also
often called simply Ravenser, was about 4 miles south of
Easington, and Old Ravenser and Sunthorpe were midway
between them. Mr Boyle believed them to have been
situated within the existing estuary of the Humber but
Mr Sheppard considers that, owing to the continual erosion
and westward shifting of the Spurn peninsula, their site is
in the sea to the east. Ravenser, under the name Hrafnseyri, is mentioned in the Orkneyinga Saga and in the
Harolds Hardrada Saga, in both cases in connection with
the battle of Stamford Bridge (1066). In the latter saga
the king "sailed from the port, called Hrafnseyri (the
raven tongue of land)." We hear nothing more of Ravenser for nearly two centuries. In inquisitions taken in
1274–1276 the men of Grimisby (Grimsby) say "that
forty years ago and more [that is about or before 1235]
by the casting up of the sea, sand and stones accumulated,
on which William de Fortibus, then earl of Albemarl,
began to build a certain town which is called Ravenesodd;
and it is an island: the sea surrounds it."

In 1251 Henry III granted a charter to William de
Fortibus to hold a weekly market and a fair of 16 days in
Ravenserot. In 1299 (reign of Edward I) Raveneserode

* See p. 53.

became a free borough and paid £300 for its royal charter, whilst Hull only paid 100 marks for its charter. In 1305 and in 1326 two burgesses represented Ravenser in Parliament. In 1310 Edward II required Ravensere to furnish one war-ship in the war against Scotland, similarly in 1314 and (Ravensrode) in 1321; in 1327 Ravenser, two war-ships. In 1332 Edward Baliol embarked at Ravenser on his expedition to claim the Scottish throne. In 1346, when Edward III laid siege to Calais, Ravenser furnished one ship and 28 mariners, whilst Hull furnished 16 ships and 466 mariners, for the expedition. The witnesses in an inquisition held in 1346 state "that two parts of the tenements and soil of the said town [Ravenserod] and more...have been thrown down and carried away" [by the sea]. The chronicler of Meaux Abbey, under date 1360, writes: "But in those days, the whole town of Ravensere Odd...was totally annihilated by the floods of the Humber and the inundations of the great sea." In the inquisition of 1401 the abbey of Meaux was shown to have lost 24 messuages and a chapel in Ravenserodde.

The first historic mention of Ravenspurn, or rather Ravenser Spurn, is in the account of the landing of Henry IV in July 1399. Edward IV landed there in 1471. The locality was practically identical with Ravenserodd. In 1822 in old Kilnsea there was a church and 30 houses. Now not a vestige of them remains. The church tower fell with a tremendous crash in 1830.

The churchyard at Withernsea having been washed away, it was decided to rebuild the church farther inland, on its present site. In 1488 the new church was completed.

In 1609 it was unroofed in a storm and it remained a ruin until 1859.

"Even since 1822, the date of the old Ordnance Survey, the village of Owthorne with a church and twelve houses has been entirely swept away, and Owthorne and Withernsea Meres have both disappeared."

Other places which have disappeared are, on the Humber: Est Somerte and Penisthorpe, and, on the sea-coast, Northorpe and Hoton in the parish of Easington, Newsom in Owthorne parish, old Waxholme, $1\frac{1}{2}$ miles N.W. of Withernsea, Monkwike N.E. of Tunstall, Monkwell and Ringbrough (except one farm house), both east of Aldbrough. Old Aldbrough has gone, so have Southorpe, Hornsea Beck (village) and another Northorpe, all near Hornsea. Withow, near Skipsea, Auburn (of which one house remains), and Wilsthorpe, 2 miles south of Bridlington, no longer exist.

Within the Humber estuary extensive reclamation of land has taken place, by surrounding areas with embankments. This has been rendered possible by the natural accumulation of "warp," the tidal alluvium deposited by the Humber. As to the origin of this silt, there has been much controversy. Some consider it to be the sediment brought down by the Ouse and Trent and other rivers; others hold that it is the clay derived from the marine erosion of the Holderness coast, swept into the estuary with the tide. Probably it is derived from both sources. The most noteworthy area of reclaimed land is Sunk Island (no longer an island) the existence of which appears to have been first recorded in the time of Charles I, when it contained about 7 acres and was separated from the

Yorkshire coast by a channel $1\frac{1}{2}$ miles wide. The progress of reclamation can be best illustrated in tabular form:

Date of Reclamation				Area in Acres	
Up to 1744	...	...	...	1,561	
1770 to 1802	...	...	...	3,289	
1826	...	...	...	1,080	
1850	...	...	...	700	
About 1867	...	...	...	1,800	
1897	...	...	...	347	
Since 1897	...	...	...	170	
					8,947

In addition:

Cherry Cob Sands in 1770	...	1,700	
Other areas, chiefly between Sunk Island and the mainland, between			
1758 and 1872 ...	...	...	1,527
			3,227
	Grand Total		12,174

Broomfleet Island lies 2 miles N.E. by N. of the confluence of Ouse and Trent. In 1866 six acres were enclosed. By 1870 sixty acres were embanked and by 1900 the channel between the island and the mainland was completely closed. By 1912 nearly 600 acres had been reclaimed.

There still remain 15,078 acres of foreshore of the East Riding in the Humber estuary; these are chiefly situated between Sunk Island and Spurn.

10. Climate.

Weather denotes the temporary atmospheric conditions of a region, whilst climate denotes the average atmospheric conditions of the region—the average weather.

The chief factors in climate are temperature (including radiation); moisture (including evaporation, humidity, precipitation and cloudiness); atmospheric pressure and wind (including storms); also the composition and chemical, optical and electrical phenomena of the air.

The climate of any place chiefly depends on (1) its latitude, (2) its position with regard to the great belts of high and low air-pressure, (3) its distance from the sea, (4) its height above sea level, and finally (5) its slope or exposure, as well as its soil and the character of its vegetation.

The East Riding lies between 53° 34′ North latitude at Spurn Head and 54° 13½′ at the Wyke, near Filey, and is in the temperate zone. The British Isles lie near the margin of the great cyclonic depression situated between Iceland and Greenland. As the wind blows from areas of high pressure to areas of low pressure and is deviated to the right in the northern hemisphere, our prevailing south-westerly winds are thus accounted for.

The climate of Britain is what is known as marine or insular; the effect of the nearness of the sea, owing to the great specific heat of water, being to moderate both the heat of summer and the cold of winter, and render the climate more equable than that of continental areas. The drift of warm water from the south-west of the North Atlantic, and the prevalence of south-west winds, increase the effect due to the presence of the sea.

Owing to the Pennine Chain intervening between the west coast of Britain and the East Riding, the vapour-laden winds from the Atlantic deposit most of their moisture before reaching the East Riding, so that its climate is, on the whole, dry. From a climatic point of view the riding may be roughly divided into two regions, that of the Wolds in the centre, with a *relatively* high rainfall and comparatively cooler temperature, and that of the low ground of Holderness and the Vales of York and Pickering, where the rainfall is less and the temperature a little higher. Differences of temperature due to latitude are very small in the riding, not more than about 1° F. Differences due to altitude are also not great, although the effect of altitude is 800 times as great as that of latitude, that is, temperature diminishes vertically far more rapidly than horizontally, measuring towards the pole. This vertical temperature gradient is variable, according to the month and locality. The Meteorological Office assumes 1° F. for 210 feet for the annual mean of maximum thermometer readings, 1° F. for 250 feet for minimum readings. The mean of these two is 1° F. for 230 feet. On this assumption, Garrowby Top, at 808 feet, would be 3·5° F. cooler than a locality at sea-level.

TABLE OF APPROXIMATE MEAN TEMPERATURE, RAINFALL, AND BRIGHT SUNSHINE.

Locality	Altitude, Feet	Temperature				Rainfall	Rain, Days	Driest Month	Sunshine, Hours
		Jan.	July	Mean Range	Annual				
Hull	12	37·5	60·2	22·7	47·8	25 7	186	Jan.	
Spurn Head	26	38·3	59·8	21·5	48·2	19·6	169	April	1273
York	56	37·7	60·4	22·7	48·1	24·8	188	Feb.	
Scarborough (North Riding).	100	38·5	59·3	20·8	47·9	26·6	191	April	1396

The wettest month, on the average, is in each case October, but the wettest and driest months vary in individual years. The low rainfall of Spurn is probably due to the absence of high land to condense the aqueous vapour. The hours of bright sunshine at York are considerably fewer than at Worthing, in Sussex, where on the average of 12 years they are 1867. The rainfall on the Wolds is several inches higher than on the low ground at their feet, but it sinks into the chalk so rapidly as to make water-supply there difficult, whilst in Holderness it is easy to obtain a supply by boring through the boulder clay into the underlying chalk.

Dr H. R. Mill gives the following details of rainfall, 1868–1902:

	Altitude (feet)	Inches	Mean computed for 35 years
Patrington	10 } 46 }	Maximum 33·64 in 1872 } Minimum 16·00 in 1887 }	24·5
Lowthorpe 1873–1902	63	Maximum 36·96 in 1876 } Minimum 19·14 in 1884 }	27·7
Old Malton (North Riding)	75	Maximum 41·79 in 1872 } Minimum 19·60 in 1887 }	26·4
Warter	230	Maximum 46·75 in 1872 } Minimum 19·10 in 1887 }	30·3

The Meteorological Office informs me that the average rainfall at Bridlington 1881–1915 is 28·5 inches.

In these five localities April is the driest and October the wettest month. Dr Mill has computed the average depth of the average annual rainfall, over the whole East Riding area, 1868–1902 as 27·4 inches, with a maximum in 1872 of 41·1 inches, and a minimum in 1887 of 18·6 inches.

From a total of 3652 observations (two per day), of winds for the five years 1911–1915, at Spurn Head, the

average number of days per year, on which a wind from
the following directions blows, has been compiled:

N.	N.N.E.	N.E.	E.N.E.	E.	E.S.E.	S.E.	S.S.E.
23·2	17·7	15·0	17·0	17·7	21·0	18·8	24·7

S.	S.S.W.	S.W.	W.S.W.	W.	W.N.W.	N.W.	N.N.W.	Calm
29·3	25·6	26·1	33·1	32·5	27·6	17·9	12·8	5·2

Average Annual Rainfall of the East Riding (*by Dr H. R. Mill*)

From the same observations it appears that gales eight,
nine, and ten of Beaufort scale (39 to 63 statute miles per
hour) occur on 9·1 days annually and fog (in the 20 years
1896–1915) on 43 days per year.

11. People—Race, Language, Settlements, Population.

For information as to the inhabitants of the East Riding in prehistoric times we have to rely on archæology. In archæology four main stages of culture and corresponding Ages have been recognised—the Palæolithic or Old Stone Age, the Neolithic or New Stone Age, the Bronze Age, the Early Iron Age.

We have no certain evidence of Palæolithic man in the East Riding. Sir John Evans records the finding at Huntow, near Bridlington, of a small pointed implement of Palæolithic type but this appears to be the only instance of such an implement in the Riding.

Palæolithic implements were only chipped but Neolithic man ground and polished such of his implements as would be thereby improved, though he used others, such as arrow-heads, merely chipped. The bow and the stone axe were his chief weapons. He inhabited caves and partly-sunk dwellings and pile-dwellings in lakes and rivers. He cultivated wheat and flax and possessed domestic animals such as the dog, sheep, goat, short-horned ox, horse, and pig. He had a knowledge of spinning and weaving flax, and probably wool, and made pottery shaped by hand, not turned in a lathe. He made canoes from tree trunks and buried his dead in long oval barrows, but sometimes in round barrows.

From numerous skeletons, the average stature of Neolithic man has been calculated as 5 feet 6 inches. His skull was long and narrow, the ratio of length to breadth (the cephalic index) being from 100:65 to 100:75, the average of 48 skulls being 100:71. Skulls are classified according to the ratio of the length, taken as 100, to the breadth. Those between 70 and 75 are called *dolichocephalic* (long-headed); those between 75 and 80 *mesaticephalic* (medium headed); and those between 80 and 85 *brachycephalic* (short-headed). It is probable that Neolithic man in Britain was of old Iberian race, of which the Basques of the Pyrenees may be considered a remnant. His descendants survive in Wales, in Scotland among the small dark Highlanders, and in south-west Ireland.

Probably about 2000 B.C. there occurred an invasion of brachycephalic people, known to archæologists as the "Beaker-folk," because they buried earthenware beakers or drinking-cups with their dead, in round barrows. Bronze came into common use soon after they arrived. Mr J. R. Mortimer measured 104 Neolithic and Bronze Age skulls (male and female) from East Yorkshire barrows (almost all round barrows) and calculated the stature of their owners from their skeletons, with the following result (corrected figures):

	Mean Cephalic Index	Mean Stature	Mean Stature of the 104
35 (33·6 °/₀) Long skulls	70·7	5 ft. 6 in.	
40 (38·5 °/₀) Intermediate Skulls	77·7	5 ft. 4·4 in.	5 ft 4·9 in.
29 (27·9 °/₀) Short Skulls	84·2	5 ft. 4·36 in.	

"It is evident that during the Round-barrow period the population was very mixed....The earliest Bronze Age invaders were no doubt racially mixed before leaving the Continent, and after their settlement here amalgamated with the aborigines, who were descended from the long-skulled people of the Long barrows." There is in the Round barrows evidence of cannibalism and "it can scarcely be questioned that it was the habit to slay at the funeral wives children and others, probably slaves."

Mr O. G. S. Crawford has brought forward evidence in support of the hypothesis that an invasion by a people using leaf-shaped bronze swords took place about 800–700 B.C. and that these invaders were the Celtic Goidels or Gaels. Their language was Gaelic, now a language spoken by the more primitive inhabitants of Ireland, Scotland, and the Isle of Man.

Probably between 400 and 300 B.C. the Celtic Brythons, from whom the name of Britain is derived, invaded the island and they may have introduced iron if it was not previously known here. The usually accepted date for the beginning of the Early Iron Age in this country is 450 B.C. The Brythonic language was later represented by Welsh, Breton, and Cornish. Mr Mortimer measured 58 skulls of men and women of the Early Iron Age (late Celtic Period) from East Yorkshire and found (corrected figures):

	Mean Cephalic Index	Mean Stature	Mean Stature of the 58
42 (72·4 %) Long Skulls	72·2	5 ft. 2·5 in.	
14 (24·1 %) Medium Skulls	77	5 ft. 3·2 in.	5 ft. 2·6 in.
2 (3·5 %) Short Skulls	81	5 ft. 1·9 in.	

About 200 B.C. the Belgæ (partly Teutonic) invaded Britain. The Roman conquest and dominion seem to have had little effect on the British races. The Angles reached Yorkshire in the fifth or sixth century. Mr Mortimer measured 61 Anglo-Saxon male and

female skulls from East Yorkshire and found (corrected figures):

	Mean Cephalic Index	Mean Stature	Mean Stature of the 61
31 (50·8 °/₀) Long Skulls	72·3	5 ft. 5·7 in.	
23 (37·7 °/₀) Medium Skulls	77	5 ft. 3·6 in.	5 ft. 4·7 in.
7 (11·4 °/₀) Short Skulls	81·1	5 ft. 4 in.	

In 867 the Danes took York and in 876 the kingdom of Deira was divided among them. Between the ninth and eleventh centuries they arrived in considerable numbers. Their invasion and conquest had more ethnological result in Yorkshire than in most other parts of England.

Considerable light is thrown upon the distribution of settlements of Angles, Norsemen, and Danes in the East Riding by its place names. Names of Celtic origin are very few. Among them are Roos (Welsh *rhos*, a moor, heath) and the river names, Hull, Humber, and Derwent.

Anglo-Saxon terminations of place names are :— *borough*, 8*, as Aldbrough ; *burn*, 9, Eastburn ; *bridge* and *ford*, 4, Stamford Bridge ; *field*, 5, Driffield ; *fleet* (a bay or stream), 7, Marfleet ; *ham* (a homestead), 38, Halsham. Of these 38 thirteen end in *ingham* (the homestead of the sons of), Riplingham, Goodmanham (Bede's Godmund-dingaham). There are also *ing* (a meadow), Sherburn Ings ; *ley* (a forest clearing), 10, Beverley ; *stead* (a place), 2, Wine-stead ; *ton* (a farm), 109, Folkton ; *well*, 5, Emswell ; *wick* (a hamlet), 21, Burstwick.

The Scandinavian place names are of more interest ; they are chiefly Danish but, to a small extent, Norse in

---

* The numbers indicate the number of recorded examples in the riding.

origin; but it is, in most cases, impossible to discriminate between these two sources. Among the most characteristic Norse terminations is *gate* (a road or path) as Huggate. As the Rev. A. Goodall remarks, "In the city of York there are many street names due to the Vikings: Goodramgate, Guthorm's road; Micklegate, the great road; Skeldergate, the shield-maker's road; Coppergate, the turner's road; Fishergate...Stonegate, formerly Steingate." Conyngstrete from O.N. *konungr*, a king, has become Coney Street.

Gill, a Norse term occurring 171 times in the West Riding has only five examples in the East Riding. The Norse *thwaite* (thveit), an enclosure, of which termination there are 68 in the West Riding, occurs twice only in the East Riding, *e.g.* Storthwaite. The predominance of Norse names in the West Riding is due to the Norsemen coming from Ireland.

Distinctively Danish terminations are *thorpe* (a hamlet), 62, Fridaythorpe; and the Danish *sten* (a stone) not necessarily as a termination, as Asteneuuic, Astene's dwelling (Elsternwick).

Of Scandinavian terminations the chief ones are *by* (a hamlet), 46, as Duggleby (Dufgal's farm); *beck* (a brook), Skirpenbeck; *carr* or *ker* (low-lying land subject to flooding), Flotmanby Carrs, Ellerker (the alder carr); *garth* (an enclosed place), Coneygarth Hill; *holme* (an island), Holme on Spalding Moor; *howe* (a hill, burying mound), Duggleby Howe; *skew* (a forest), Heslesskew; *toft* (a homestead), 3, Willitoft (the toft of the willows); *wick* (O.N. vík, a bay), Thornwick, Selwicks, Sanwick; *with* (a wood), 5, Skipwith.

Collecting from the Domesday record every name distinctively Danish, 181 in all, the Rev. A. Goodall finds in North Riding (west of Northallerton) 19; (east of Northallerton) 40; in the West Riding (north of River Aire) 26; (south of River Aire) 23; in the East Riding 73; showing "how strong is the evidence that East Yorkshire is more Danish than West Yorkshire."

Mr G. G. Chisholm has pointed out that in Holderness the alluvium is usually avoided in choosing village sites, with the notable exception of the city of Hull, which he regards as an artificial creation. Hessle, Cottingham, Sutton, Hedon, Keyingham, Ottringham, Patrington, Welwick, Skeffling, and Easington are on boulder clay near its southern margin. The influence of springs in determining the sites of villages has been described in Chapter 5.

The population of the East Riding in 1801 was 111,192, including 29,848 in Hull. In 1921 the population of the administrative county of the East Riding and associated city and county borough of Kingston-upon-Hull was 460,880 or 393 per square mile. This number per square mile is very misleading, as Hull, with an area of 9042 acres, had a population of 287,150 or 62·3 per cent. of the whole. If we deduct the area and population of Hull, the administrative county proper has an area of 1158 square miles and a population of 173,730 or 150 per square mile.

12. Agriculture.

The East Riding is pre-eminently an agricultural county, though, if we include Hull, the majority of the population is employed in other occupations. The number of persons employed in agriculture, in the administrative county and in Hull in 1921, was 23,243, of whom 19,954 men and 600 women were farmers, graziers, and farm workers; 2194 men and 101 women were gardeners, nurserymen, seedsmen and florists; and the remaining 394 men were woodmen, agricultural machine-owners or attendants, etc.

The number of agricultural holdings in the riding in 1922 was 6932, the total acreage of the holdings being 667,117 acres, and that of the average holding 96 acres. The average holding in England and Wales is 62 acres, therefore the East Riding is a county of large holdings.

Of the 667,117 acres, 455,571 are arable land, and 211,546 acres permanent grass (the latter not including 8863 acres of mountain and heath land used for grazing), the proportion of arable to permanent grass being 68 per cent. arable to 32 per cent. pasture. It is evident that the East Riding is pre-eminently a corn or crop-growing, rather than a pastoral or stock-breeding, county. The area under the various crops is shown in the diagrams on pp. 172–174.

Compared with the areas under cereals the space devoted to beans (10,578 acres) and peas (6611 acres), seems small. Potatoes occupy 14,865 acres and mangold 7201 acres.

About 1770 the Wolds formed one unenclosed sheep-walk, over which you might gallop in all directions, without being troubled by a fence. Sir Christopher Sykes was the

first to lay out plantations on the Wolds and his plantations, which ultimately reached about 2000 acres, not
only afforded shelter to the cattle and to the growing
corn but also furnished materials for fencing. Between
1771 and 1801 "by his assiduity and perseverance in
building, planting, and enclosing on the Yorkshire Wolds"
he "caused what was once a bleak and barren tract of
country to become now one of the most productive and
best cultivated districts in the County of York." On the
Wolds the growing of corn is connected with the breeding
and feeding of sheep, mainly Leicesters, which eat the
clover and most of the turnips and swedes on the land,
and, by their treading and droppings, enrich it for the
corn-growing. Sir Tatton Sykes (1772–1863) was the
first to introduce bone manure for growing turnips. He
noticed that on the patches of grass on which his fox-
hounds were fed the vegetation was much thicker and
more luxuriant than elsewhere, and he forthwith tried the
experiment of crushing bones for manure. The usual
rotation on the Wolds is (1) Turnips, Swedes, etc.,
(2) Barley, (3) Clover, (4) Wheat or Oats.

While water-supply is a difficulty on the Wolds, the
reverse obtains in Holderness, the Vale of Pickering, and
the Vale of York, where drainage is of first-rate importance. In Holderness the Commission of Sewers for the
East Parts of the East Riding of the County of York has
jurisdiction over 228,330 acres, but within and to be sub-
tracted from that area the Beverley and Barmston Drainage
Commission, constituted 1798 and 1880, has jurisdiction
over 2182 acres at the Sea End (north of the Barrier)
discharging at Barmston Drain, 5¼ miles south of Brid-

lington and over 10,413 acres south of the Barrier, discharging into the River Hull above Scott Street Bridge, in Hull city. The Holderness Drainage Trustees have a taxable area of 11,515 acres. Other drainage commissioners have jurisdiction in districts adjoining Beverley and Skidby, Hessle and Anlaby, Cottingham, Keyingham, Ottringham, Thorngumbald, and Winestead, all to be deducted from the 228,330 acres. In the Holderness district wheat and oats occupy about equal areas and each predominates over barley. Lincoln sheep are preferred to Leicesters.

In the Vale of York Drainage Commissioners control districts of Ouse and Derwent, Bishopsoil, Wilberfoss, and Market Weighton. The alluvial clays are chiefly under permanent grass. The sandy soils are most suitable for oats, rye, and potatoes, and the blown sand for carrots. On the warp a rotation of (1) bare fallow, (2) mustard (for seed), (3) mangolds, (4) wheat, (5) clover or beans, (6) wheat, is often adopted.

In the Vale of Pickering an Act was obtained in 1800 for draining the eastern end, as far as Yedingham Bridge. A portion of the flood water was turned into a New Cut, connecting with Scalby Beck (North Riding), and a new river or drain was cut from Muston to Yedingham, draining about 12,000 acres. In 1846 the Rye and Derwent Drainage Act gave powers to remove obstructive mill-dams, locks, etc. on the Rye and Derwent. In the vale, barley predominates over oats and oats over wheat. Sheep are fewer per hundred acres, but cattle more numerous, than on the Wolds. Some additional details of agricultural interest will be found in Chapter 14.

13. Industries.

The industries of the East Riding are chiefly concentrated in the city of Hull. In almost every industry there are more workers in Hull than in the whole of the administrative county. The numbers of persons employed in each industry are taken from the census of 1911.

In General Engineering and Machine Making 6387 men were employed in Hull. They include iron founders, blacksmiths, erectors, fitters, turners and others. Marine engines and boilers have been constructed there since the last decade of the eighteenth century, and it is stated that two firms between them have already constructed over 2000 sets of engines and boilers for steam trawlers. Power-driven screens for cleaning seeds, seed-decorticators, hydraulic presses for seed-crushing and oil expression, cattle feed-cake making machines, chemical solvent oil-extraction plant, bone-crushing mills, dredges, grabs, and excavators, cement mill machinery and paint mill machinery are made in Hull. In the administrative county 1221 men were employed in Engineering (including agricultural machinery making) at Great Driffield, Beverley, Market Weighton, Bridlington, Pocklington, etc. In York 1238 men were similarly employed. Tin-plate goods makers employed 487 men and 989 women in Hull.

In Ship and Boat Building 2767 men were employed in Hull and 580 in the administrative county. As mentioned in Chapter 9, Hull furnished in 1346, 16 ships for the siege of Calais and these were probably built locally. When the Royal Navy was built of oak, many warships were built at Hull (14 between 1739 and 1815), Hedon, Paull, and Hessle Cliff. In 1693 an 80-gun ship was launched at Hessle Cliff. Iron- and later, steel-ship-building has been carried on in Hull for more than seventy years. A ton of shipping is 100 cubic feet of internal space. In 1921 the s.s. *Tekoa* of 9740 tons gross, with engines of 6100 i.h.p. and three other steamers, of which the total gross tonnage was 10,373, were launched at Hull; four coasting steamers, totalling 1935 tons, at Beverley; at Hessle, a vessel of 2829 gross tons, the largest ever built there, as well as another of 254 tons. In the ten years 1910 to 1919 there were built at Hull 143 sailing vessels, averaging 92 net tons and 299 steam vessels, averaging 423 net tons. Owing to the prevailing depression, Selby (West Riding) and Beverley, which usually build between them about 80 steam trawlers annually, built none in 1921. Barges are built at Beverley and boats at Bridlington, Filey, Flamborough, Hornsea, and Howden.

House-building employed 6053 men in Hull (brick-layers, masons, joiners, painters, plumbers, etc.) and 3530 in the administrative county. In making Furniture and Fittings and Upholstery 824 men and 183 women were employed in Hull and 246 men in the administrative county. Other workers in wood and bark, such as sawyers, wood-cutting machinists, coopers, etc., numbered 2565 men and 59 women in Hull, and 172 men in the administrative county. In Plaster and Cement Manufacture, chiefly the latter, 373 men were employed in Hull, and in brick and tile manufacture 246 men in the administrative county. Cement is made in Hull and large Portland cement works, employing additional men, with an output of over

2000 tons per week, have been recently established at Melton, where the necessary chalk is dug by a steam navvy and mixed with the warp from near the Humber shore.

In the manufacture of Oil, Grease, Soap and Resin, etc., 2764 men and 55 women were employed in Hull and 349 men in the administrative county (Great Driffield and Beverley). This trade of Seed-Crushing, Oil Extraction and Refining, may be regarded as the staple trade of Hull, which is the chief seat of the industry in the country. It was connected with, and is the successor of, the refining of Whale Oil (see Chapter 15, p. 75). Seeds are crushed in hydraulic presses or, if the residual cake is not required for cattle-feeding, the oil is extracted by a chemical solvent. Among the seeds imported into Hull for this purpose in 1920 were Castor, from Bombay and Brazil 11,207 tons; Cotton, 223,622 tons; Flax or Linseed 168,452 tons from the Argentine and India ; Rape 20,136 tons; Soya Beans, *Glycine hispida*, 13,535 tons from China, Manchuria and Japan; Palm Kernels (of the Oil Palm, *Elæis guineensis*, of West Africa) 39,297 tons, all other kinds 18,829 tons. From the palm kernel oil are made margarine, lard, and soap. Coconut oil and cotton seed oil are also used in margarine. Feeding-cakes for cattle are made from the residue of palm kernels, linseed, rape seed, and cotton seed. Linseed oil is used in paints, putty, linoleum, and soft soap ; Soya bean oil in soap and for edible purposes.

In making Dyes, Paint, Ink, and Blacking and in Manufacturing Chemical works, 1563 men and 1887 women were employed in Hull and 60 men in the administrative county. This group includes the manufacture of oil paints, varnishes, naphthaline, carbolic acid, black lead, coal tar products, starch and laundry blue.

In Tanning 433 men were employed in Hull and 562 in the administrative county, *e.g.* Beverley. In York 141 persons were employed on skins and leather (not boots and saddlery) and there is a tannery.

In Canvas, Sailcloth, Sacking, and Net manufacture 64 men and 423 women were employed in Hull and in textile manufactures 91 men and 81 women in the administrative county. Hull has sail-lofts and rope-works and makes tarpaulins, covers, and awnings.

Owing to the great quantities of grain and flour imported into Hull, valued at 23 million pounds in 1920, it has become one of the chief milling centres of the world. It has splendid modern flour mills and employed 655 men as millers and cereal food makers ; in the administrative county the numbers are not specified, but millers occur in 26 localities. In York 307 men were similarly employed and, in addition, 1134 men and 2043 women were engaged in Chocolate and Cocoa making there.

Makers of Spirituous Drinks (chiefly brewers and maltsters) employed 201 men in Hull and 120 in the administrative county, *e.g.* Aldbrough.

Railway, Shipping, and Fishery employees are mentioned in the chapters devoted to those subjects.

14. Minerals.

There are no mines in the East Riding and the district is not rich in mineral wealth. Under the Quarries Act of 1894, only the output of quarries over 20 feet deep is recorded. Of such quarries 28 were at work in 1920 in the riding, and 129 men were employed inside and 66 outside the quarries. Large unrecorded quantities of material must be obtained from shallow quarries. The values are not separately given for the East Riding, but I have calculated them at the average prices shown for the United Kingdom.

Chalk. The output of chalk in 1903 was 51,297 tons; in 1913—174,141 tons; in 1920—305,611 tons, value £36,253. Paradoxical though it may appear, the soil of the Wolds, though often less than six inches thick and immediately overlying the Chalk, contains from 62 to 74 per cent. of siliceous sand and only three to five per cent. of carbonate of lime. It is therefore necessary periodically to marl, or rather chalk, the fields to prevent turnips from running to "fingers and toes" (a fungoid disease). Shallow pits are opened in the middle of the fields and the chalk is spread over the surface in the autumn, so that the winter frosts may break up the lumps. Deep chalk-pits exist at Hessle and other localities on the Wolds. The chalk is used for ballast, for making lime and whiting (used for distemper, plastering, putty making, and for producing carbonic acid gas for aerated waters), and to a decreasing extent for road repairs. It is too soft for road metal but, when mixed with tougher materials, makes the roads bind well. As mentioned in the chapter on Industries, it is excavated in large quantities at Melton and elsewhere for making Portland cement. A great part of the walls of Watton Priory was built of chalk.

Clay. The clay dug in deep clay-pits in 1920 amounted to 21,532 tons, value £7112. Keuper Marl is dug at

Bishop Wilton for brick-making and, south of Holme on Spalding Moor, for marling the sandy flats of the district. About Houghton (south of Market Weighton) the Lower Lias is entirely masked by a thick covering of blown sand. "This light soil is often improved by marling with the Lias clay below, so that land which was formerly covered with ling and furze, was, in 1848, worth 20*s.* an acre." Boulder clay is abundant in Holderness, but if it contains chalk fragments (or pyrites) it makes bad bricks, for, in baking, the chalk is converted into quicklime, which, when the bricks are wetted, causes them to split. Where the clay contained originally little chalk or has been naturally decalcified, it gives better results. The Humber warp makes excellent bricks, tiles, and drain pipes and, when mixed with chalk, Portland cement.

Flint and Chert. Flint is of course obtained from the Chalk and is used, to some extent, for road metal. The output in 1920 was 740 tons, value £309.

Gravel and Sand are usually obtained from shallow gravel pits, which accounts for the small quantity of 4350 tons (value £839) recorded. The so-called Inter-glacial Marine Gravels of Kelsey Hill, Burstwick and other localities have yielded large supplies, and sand is abundant in the Vales of York and Pickering.

Gypsum, Anhydrite, and Salt. Fibrous gypsum occurring in the Keuper Marl was formerly worked in Scrayingham. Grey anhydrite was met with in the Permian at 2879 feet in the Market Weighton boring and a thickness of 221 feet was proved (end of boring). In the Middle Permian Marl in the same boring a bed of rock salt, 23 feet thick, occurred at 2482 feet.

Ironstone. The Dogger, at the base of the Lower Oolites, is 12 feet thick at Kirkham, where it has been worked as an ironstone, but apparently without much success.

Limestone (other than Chalk). In 1920—17,983 tons of limestone (value £5634) were quarried. Some of the thin beds, such as the limestone bands of the Lower Lias south of Garrowby, the Millepore Bed at Westow, and the Hydraulic Limestone east of Hotham, have been quarried for road metal. In south Yorkshire about Cave, the Cave Oolite, a soft sandy oolitic limestone, which is the equivalent of the Millepore Bed, is the principal source of lime. It is used along the Wold foot between Sancton and Brough. It has been quarried at Brantingham, Newbald, and Brough and has been used for interior work under the name "Cave Marble." The thirteenth century work at Beverley Minster was of Newbald stone and it was used in the monasteries of Holderness and, more recently, in the construction of the Hull docks. The other Jurassic limestones have been burnt for lime, the principal one being the Upper Limestone of the Corallian, *e.g.* of the west quarry at North Grimston. The overlying North Grimston Cement Stone is a close-grained argillaceous limestone, occurring between the villages of Birdsall, North Grimston, and Langton. "It forms a lime which, from its similarity to that obtained from the Lias of the south of England, has been called Blue Lias Lime or Lias Hydraulic Lime." The cement made from it has been employed in important buildings at Scarborough and elsewhere.

Sandstone. Though no sandstone is recorded as having been quarried recently in the East Riding, the Lower

Calcareous Grit has been worked at several localities, *e.g.*
Birdsall Quarry, for important buildings. Ralph de Nevill
granted stone from the quarry at Filey to the Canons of
Bridlington, probably between 1194 and 1220, for building
their monastery and offices. This stone was probably the
Jurassic grit of Filey Brig.

Phosphatic Nodules (the Coprolite-Bed of Judd) occur at
the base of the Speeton Clay and were formerly mined at
New Closes Cliff.

15. Fisheries.

Before describing the modern fisheries of the East Riding
a few words may be devoted to the Whale Fishery, for
which Hull was long famous. The Greenland Whale
(*Balæna mysticetus*), the objective of the whaling voyages,
is not a fish but a cetacean mammal. It was in 1598 that
Hull merchants first fitted out ships for the whale fishery.
A government bounty of between 20s. and 40s. per ton,
on the tonnage of whaling ships of 200 tons and upwards,
was paid for many years. Vessels of about 350 tons were
considered to be most suitable. During the period of
80 years (1772 to 1852), 194 different ships sailed from
Hull to the whale fisheries of Greenland and Davis Straits,
many of them, *e.g.* the *Truelove* 58, making numerous
voyages in that period. In each of the years 1818 and
1819 sixty-four whalers sailed, the largest number ever
sent out from Hull. During the 80 years, the Hull whalers
brought home 171,907 tons of oil, value £5,158,080, an
average of about £30 per ton and of 88 tons of oil per ship,
per annum. The gross amount of whalebone obtained was

8556 tons, which realised £1,691,000, an average of nearly £200 per ton. After 1846 the trade gradually decreased, whales became scarce and, owing to the introduction of gas lighting, oil fell to so low a price as to be unremunerative. The *Truelove*, the last of the sailing whalers, sailed from Hull in 1868.

The means employed for capturing sea-fish are nets and lines and, of these, nets are by far the more important. Among nets the trawl takes the chief place. It is flattened conical in form, about 100 feet long, with a very wide mouth in front and a very narrow opening at the "cod or purse" end, behind. When in use the cod end of the net is closed by a draw-rope or cod-line. When, usually after six hours, the trawl with its catch is hauled aboard, the cod-line is untied and the contents of the net fall on the deck. On steam trawlers the Otter Trawl is used. It has at each side of the wide mouth, a trawl board of deal or elm, 8 to 10 feet long, 4 to 5 feet high, about 3 inches thick and shod with iron. These are so attached to the net and the wire rope warps, which haul the trawl, that they stand on their edges and act as does a kite in air, to keep the mouth of the net widely open. The chief fishes caught in the trawl are plaice, sole, cod, haddock, skate, hake, ling, and gurnard. Long-line fishing is practised between Bridlington and Spurn Head. Scale-models of Beam- and Otter-trawls and Drift-nets, and many whaling relics, are shown in the Museum of Fisheries and Shipping in Hull. British steam trawlers carry on their work in fishing grounds as far distant as the White Sea, Barents Sea, and Iceland on the north, and the coast of Morocco on the south, but the total weight of fish caught in the North Sea is greater

than that from any of the other localities. Fishing has no
doubt been carried on by the fishermen of the East Riding
from very ancient times, but it is only since the forties of
the nineteenth century, that deep-sea fishing, with the
allied industries of ice-manufacture, drying, curing, box-
making, and the exportation of salt-fish, has attained great
importance. In 1865 about 300 sailing trawlers belonged
to the port of Hull and their usual tonnage was about
80 tons. One of the first steam trawlers to be built was
launched at Hull so lately as 1881, and the use of steam
has enormously increased the available area of fishing
grounds and the take of fish. In 1921 Hull owned 241
steam trawlers, varying in gross tonnage between 139 and
351 tons, with an average of 241 tons, the total tonnage
being about 58,000 tons. In 1922 the vessels fishing from
Hull numbered 268 steam trawlers and 36 seine-net boats,
and 235 of these steam trawlers were manned by 10 men
each or 2350 men and 150 boys. There are no drifters
employed in herring fishing. The number and tonnage of
the fishing boats of the other East Riding localities are
given in the following table—

	Motor Cobles	Gross Tonnage of each boat	Sailing Cobles	Gross Tonnage of each boat
Filey . . .	14	15	11	3
Flamborough .	6	7	38	3
Bridlington . .	37	16	4	4
Hornsea . .			2	2½
Aldbrough . .			2	3
Withernsea . .	1	5	3	2
Easington . .			4	2
Paull . . .			4	7
Total	58	849 tons	68	216 tons

The motor cobles have a crew of 3 to 5 men and the sailing cobles 2 to 3 men. The number of East Riding fishermen, exclusive of Hull, in 1922 was 463 men and 15 boys. The wet fish landed at Hull in 1919 amounted to 1,177,673 cwt., value £2,440,576, compared with 2,627,764 cwt., value £6,772,426, landed at Grimsby, our greatest fishing port.

The wet fish landed at Hull in 1919 arranged in order of importance (total weight) beginning with the most abundant were :—Haddock, Cod, Plaice, Whiting, Catfish, Ling, Mackerel, Dabs, Gurnard, Halibut, Skates and Rays, Turbot, Lemon Soles, Herrings, and Witch Soles.

Other fish were 6056 cwt. Only 164 cwt. of true soles were landed at Hull. At Filey the total wet fish landed weighed 1126 cwt., including 925 cwt. cod and 175 cwt. haddock ; at Flamborough total 2339 cwt., including 2289 cwt. of cod ; at Bridlington 12,966 cwt., including 10,055 cwt. cod, 1207 cwt. herrings, 983 cwt. plaice ; at Hornsea 6 cwt. ; at Withernsea 123 cwt. Crabs landed at Filey numbered 62,304 (and 9186 lobsters), at Flamborough 471,600 (and 11,895 lobsters), at Bridlington 90,360 crabs ; at Hornsea 11,470, at Withernsea 50,680. The total value of wet fish and all shell fish landed at Filey was £5027, Flamborough £12,881, Bridlington £37,398, Hornsea £383, Withernsea £956.

16. Shipping and Trade.

At the present time Hull is the only important port in the East Riding, though, as mentioned in Chapter 15, Filey, Flamborough, Bridlington Quay, Hornsea, Ald-

brough, Withernsea, Easington, and Paull possess fishing
vessels and, as mentioned in Chapter 8, Bridlington has a
harbour. In the past Hull was by no means always without a rival. The rapid rise and equally rapid decline of
Ravenser and Ravenser Odd, between 1235 and 1360,
have been described in Chapter 9.

Hedon (*pron.* Heddon) was, in medieval times, a port
of some importance, though even in the reign of King
John it was less wealthy than Hull, for, in 1203-4, a tax
of one-fifteenth on the goods of merchants produced
£344. 14*s.* 7*d.* in Hull, but only £60. 8*s.* 4*d.* in Hedon.
Leland, writing between 1535 and 1543, says of Hedon
" it is evident to see that sum places wher the shippes lay
be over growen with flagges and reades....Treuth is that
when Hulle began to flourish Heddon decaied." Poulson
(1841) says " the Old Harbour which insulated the town,
consisting of about 300 acres, where in ·the reign of
Edward III lay vessels of superior size...is now luxuriant
meadow." When Thurstan was archbishop of York
(1114—1140), he encouraged the merchants of Beverley to
connect their town with the River Hull, by a canal called
Beverley Beck, though it does not seem clear whether the
work was executed at that time. It certainly existed in
1344. In 1269 the River Hull was made navigable for
ships between Hull and Beverley. In the fourteenth century Beverley was a more important town than Hull, for,
from the subsidy roll of 1377, it appears that, in Beverley,
the number of persons returned above the age of fourteen
was 2663, whilst the number above the same age taxed in
Hull was only 1557. The entire population of Beverley
at that period has been calculated as being about 4000 ;

that of Hull 2336. The dissolution of religious houses and
the transfer of the staple to Hull so disastrously affected
Beverley that in 1599 there were 400 tenements and
dwelling houses uninhabited there. The more advan-
tageous position of Hull enabled it, in course of time, to
completely surpass Beverley in its foreign trade. The little
town of Wyk, the ancestor of Hull, at first only afforded
refuge or shelter in the River Hull. Later it became a
port of transhipment for goods, which could be taken by
water to the limits of navigation, on the Trent at Burton,
the Idle at Bawtry, the Don at Doncaster, the Aire at
Ferrybridge, the Wharfe at Tadcaster, the Ouse and Ure
at Boroughbridge. With the advent of canals, *e.g.* the
Aire and Calder Navigation in 1698, the Leeds and Liver-
pool Navigation in 1770, the importance of Hull as a
transhipment port increased. Lastly the need of docks for
merchant ships of continually increasing size led to the
construction of many docks in Hull for which the deep
water approach to the port peculiarly fitted it. Until 1775
the whole of the commerce of Hull, as regards wharves and
quays, was confined to that part of the River Hull known
as the Old Harbour, but in 1773–4 the Hull Dock Company was formed, and constructed the first dock in Hull
(and one of the earliest in England), now called the
Queen's Dock, present area (to nearest acre), $10\frac{1}{4}$ acres,
opened in 1778. It communicates with the River Hull.
The second dock to be completed was the Humber Dock,
$9\frac{1}{2}$ acres, opened 1809, communicating with the Humber
on the south. The third dock, Prince's (or Junction) Dock,
6 acres, was completed in 1829. These three docks occupy the site of the ancient city walls and moat and, in

conjunction with the rivers Hull and Humber, they completely surround the ancient city. The Railway Dock,
2¾ acres, opened 1846, lies at right angles to the Humber
Dock. The Victoria Dock, 25½ acres, opened 1850, with
Timber Ponds 24 acres, lies on the east side of the River
Hull. In 1869 the Albert Dock, west of the Humber
Dock and parallel with the Humber shore, was opened
and this was followed, in 1880, by the William Wright
Dock on the west; the area of the two combined is
31 acres. In 1883 St Andrew's Dock (including extension), 20 acres, west of William Wright Dock, was
opened. In 1885 the Alexandra Dock, 46½ acres (extension in 1900, seven acres) east of Victoria Dock, was
completed, and in 1914 King George Dock, 53 acres
east of Alexandra Dock, was opened by King George.
The Riverside Quay of the North Eastern Railway is
2500 feet long. The total dock water area amounts to
235½ acres. In addition there are 12 dry or graving docks
and 9 slipways. The Saltend Oil Jetty, opened 1914, at
Salt End ¾ mile N.N.W. of Paull, extends 1500 feet into
the Humber and has 30 feet of water at L.W.O.S.T.
In 1921 there were employed in Hull in the merchant
service including ship-owners 5929 men; as bargemen
and boatmen 1220 men, and as dock employees 7729
men.

As previously mentioned, Hull in 1346 furnished 16
ships and 466 mariners for the siege of Calais. In 1701–2
the merchant ships belonging to Hull were 115 in number,
with an estimated tonnage of 7564 tons. In 1921 the
steamers owned in Hull numbered 122 with a gross tonnage from 116 to 7542, total tonnage (of 121) 259,397.

King George Dock with electric coal conveyor, Hull

Of these 18 were above 4000 tons and 37 between 2000 and 4000 tons, and this total tonnage does not include the 241 steam trawlers, with a tonnage of 58,000, nor the 50 or more tug boats, nor, of course, the sailing ships. Net registered tonnage is usually about two-thirds of gross

TRADE OF HULL

Chief Imports 1920	Value in millions of pounds
Chemicals, Drugs, Dyes and Colours	2·0
Cotton, Raw and Waste .	1 5
Fruit (fresh)	2·3
Butter	1·8
Margarine	2·0
Potatoes, Onions, raw vegetables	1·7
Wheat	15·8
Barley, Oats, Maize, Rice, dried Peas and Beans, Wheat Flour . . .	7·1
Meat, including Bacon . .	3·8
Iron and Steel and Manufactures thereof . . .	2·1
Petroleum	2·7
Paper, cardboard, paper making materials . . .	5·0
Seeds and nuts for oil . .	14·5
Flax, hemp, and tow . .	1·1
Wood and timber, unmanufactured	7·9
Wool	7·0

Total Imports of Hull £95,577,000

Chief Exports 1920 Of British and Irish produce and manufacture	Value in millions of pounds
Apparel (garments, boots, hats, hosiery) . . .	2·9
Chemicals, Drugs, Dyes and Colours	1·7
Coal	1·2
Coke	1·1
Cotton Yarns . . .	5·1
Cotton manufactures . .	12·4
Machinery	3·8
Iron and Steel and .Manufactures thereof . . .	4·0
Rubber manufactures . .	1·0
Oils, fats, and greases . .	5·7
Wool (raw) waste and noils .	2·3
Woollen and worsted yarns and manufactures . .	16·4

Total Exports of British produce £71,417,000
Total Exports of Foreign and Colonial Merchandise £5,958,000
Total Exports of every kind £77,375,000

The coal exported in 1920 amounted to 328,000 tons, and 74,000 tons coastwise.

In 1913 coal exported amounted to 4,629,000 tons, and 933,000 tons coastwise.

tonnage. The highest total of shipping registered (not necessarily owned) in Hull was in 1914, when 584 sailing vessels, tonnage 46,705 and 692 steamers, tonnage 248,006, were recorded ; in 1920 the sailing vessels numbered 467, tonnage 42,401, steamers 521, tonnage 188,169. The combined inward and outward tonnage of Hull's foreign trade reached its highest level, 8,259,538 tons, in 1913 and in the same year its coastwise trade amounted to 2,644,527 tons. In the value of its trade Hull is the third port of the United Kingdom. The value of the total imports and exports of Hull gradually increased from 66 to 90 million pounds in the ten years 1909 to 1918, but in 1919 it suddenly rose to 148 and in 1920 to 173 millions, probably owing to inflated prices. The chief imports and exports entering and leaving Hull in 1920 are shewn in the table on page 83.

17. History.

It was not until A.D. 43 that the Roman Emperor Claudius commissioned Aulus Plautius to undertake the conquest of Britain. By the year 47, the country south and east of a line joining the Severn and the Wash was subdued. Aulus Plautius was succeeded in 47 by P. Ostorius Scapula, who defeated the Brigantes in 49. The Brigantian territory appears to have extended from the estuaries of the Humber and Dee to the isthmus between the Tyne and Solway Firth.

According to Ptolemy, the geographer of Alexandria, who flourished in the first half of the second century A.D., the region now known as the East Riding was inhabited by the Parisii, possibly a branch of the Parisii on the Seine, whose name is preserved in that of the city of Paris. The Parisii of Britain were independent of the Brigantes. In the year 78 Julius Agricola, who had spent much of his life in Britain, was appointed governor, and, according to the historian Tacitus, who was his son-in-law, subdued the Brigantes in 79 and we may reasonably assume that the Parisii would not retain their independence. Agricola was recalled to Rome in 85, and shortly afterwards the Brigantes rebelled. About 115 or 120 the Northern Britons again rose in revolt and destroyed the Ninth Legion, posted at York. In 122 the Emperor Hadrian came to Britain, bringing the Sixth Legion to replace the Ninth. He suppressed the revolt and ordered the construction of a wall of stone, superseding one of turf, between Tyne and Solway, not only as a barrier against the Caledonians, but also as a means of keeping the Brigantes under control by its garrison. The Brigantes were not finally subdued until the reign of Antoninus Pius (138–161). In the year 208 the Emperor Septimius Severus visited Britain, and led his troops in two campaigns against the southern Picts and the Caledonians, returning worn out to Eburacum (York), where he died in 211, after causing the wall of Hadrian to be reconstructed. The Emperor Constantius Chlorus spent much time in Britain, and also died at Eburacum in 306, and his son Constantine the Great was proclaimed there. He was the first emperor who caused Christianity to be recognised by the state. From 360 to 367 the attacks of the Picts, Scots and allied tribes, and of the Saxons, became more frequent and serious, and in the latter year they devastated a great part of Britain, though driven back and defeated by Theodosius in 368, when the Roman rule was re-established. Raids and attacks were, however, soon renewed, and in 410 the Emperor Honorius was in such difficulties that he bade the Britons defend themselves against the barbarians, which they did, at the same time establishing a government of their own. The Roman rule in Britain thus came to an end.

Little appears to be known as to what happened in Yorkshire during the next century and a half, but by the end of that time, the Angles had established two kingdoms north of the Humber—Deira, which extended from the Humber to the Tyne, and Bernicia, which lay between the Tyne and the Forth. These kingdoms were frequently united in the kingdom of Northumbria. Edwin, king of Northumbria, with his Witan, was converted to Christianity by Paulinus in 627. His pagan priest Coifi, leaping on horseback, profaned the sacred temple of Goodmanham, by hurling a spear into it and bade his companions pull down and burn the sanctuary and its buildings. In the course of the eighth century fifteen kings reigned in Northumbria "and of these five were deposed, five murdered, two voluntarily abdicated the throne." In 827 Northumbria acknowledged the supremacy of Egbert of Wessex, the first king of all England. In 867 the Danes, under Healfdene and Inguar, rode northward from East Anglia, took York and in 868 defeated and slew Osbert and Ælla, the rival Northumbrian kings. In 876 Healfdene partitioned Deira among his followers. Danish kings ruled in York. In 927 Athelstan, king of England, drove out Guthfrith, king of York, gained in 937 a decisive victory over the Yorkshire and Irish Danes and the Scots at Brunansburh (a locality not as yet identified, possibly Birrenswark, in Dumfriesshire) and maintained his rule over Yorkshire until his death in 940. Between 940 and 954 Northumbria was ruled by an alternating series of Danish and English kings, but in 954, the Northumbrian king Eric, son of

Harold Blue-tooth of Denmark, was slain and Northumbrian royalty came to an end. The English king, Edred, constituted Northumbria an earldom, with its seat at York. In 993 both banks of the Humber were ravaged by the Danes. In 1013 Sweyn of Denmark appeared with his fleet in the Humber and Northumbria submitted to him. In 1016 his son Cnut (Canute) became king of England and was succeeded by his sons, Harold Harefoot in 1035 and Harthacnut in 1040. Edward the Confessor, son of Etheldred the Redeless, succeeded in 1042 and died in 1066. In 1065 Tostig, earl of Northumbria, was outlawed by the thegns of his earldom on account of his frequent absence and his mercilessness. He returned in 1066, in alliance with Harald Hardrada, king of Norway. They sailed up the Ouse and left their ships at Riccall. The earls Edwin of Northumbria, and Morkere of Mercia gave battle to them at Fulford, 1½ miles S. of York, but were defeated. York consequently offered no resistance. Hearing of the approach of Harold, king of England, Tostig and his allies marched to Stamford Bridge, where they concentrated on the east bank of the Derwent and held a bridge-head on the west bank. On September 25 Harold of England, at the head of 60,000 men, attacked them and gained an overwhelming victory. Tostig was killed by Harold, his brother, and Harald Hardrada was slain by an arrow. Nineteen days later Harold was himself defeated and killed, at the battle of Hastings, by the army of William the Conqueror.

Several years passed before the conquest of England was completed. Even by the summer of 1068 no step had been taken practically to subjugate Yorkshire, but in that year

York opened its gates on the approach of the Conqueror. He at once built a castle between the Ouse and Foss to overawe the city, but in January 1069 the citizens of York revolted, killed Robert Fitz-Richard and many of his companions, and laid siege to the castle. William came, slaughtered and dispersed the besiegers, and built a second castle on the opposite bank of the Ouse. In September 1069 the English, assisted by a Danish fleet, attacked and took the castles of York, sparing but few of their defenders. They then dispersed. William returned to the city, much of which, including the cathedral, had been burnt, and ordered the castles to be repaired. He then exacted a terrible vengeance, by relentlessly harrying Yorkshire and the neighbouring shires. All who withstood him were slain, houses and their contents and property of every kind were burnt and destroyed. Thurstan, archbishop of York (1119 to 1139) granted to Beverley town its first charter, making it a free borough and granting its burgesses the right to have a merchant guild and freedom from toll throughout Yorkshire. In 1293 Edward I obtained by exchange, from the Abbey of Meaux, the adjacent vills of Wyke and Myton and caused the name of Wyke, first mentioned about 1160, to be changed to Kingston-upon-Hull. In 1299 he granted Kingston a charter, constituting it a free borough, with two markets per week and an annual fair. The town was represented in Parliament in 1305. In 1322 the Scots under Robert Bruce, after defeating Edward II at Byland, plundered all the East Riding. In 1340 the Commons granted Edward III 30,000 sacks of wool. It must not be supposed that the tax was paid in kind ; the value of a sack of wool was about £4 in money of the time. The relative wealth of counties is shown by the fact that the East Riding was assessed at 500 sacks, York 50, the West Riding 334, the North Riding 275, Lancashire 256, Norfolk 2207. In 1349 the plague known as the Black Death was at its height in Yorkshire. In the East Riding only sixty of the ninety-five parish priests escaped. The Abbey of Meaux lost four-fifths of its fifty monks. Henry VI in 1440 granted a charter by which Kingston-upon-Hull was erected into a county.

In 1536 the dissolution of the monasteries and other unpopular measures of Henry VIII caused a rebellion in Lincolnshire, Yorkshire, and Durham known as the "Pilgrimage of Grace." Its leader was a lawyer named Robert Aske, who, with 35,000 men, confronted the King's forces of 8000 under the Duke of Norfolk, on the opposite side of the river Don at Doncaster. A conference was held, the King's pardon and a parliament at York were promised, and the rebels dispersed; but when another rebellion occurred the following year, the King made it an excuse for executing Aske (and others), though Aske had done his best to discourage the rising.

In 1569 a rising took place in Yorkshire and Durham, with the object of releasing Mary Queen of Scots from Tutbury Castle and restoring the Roman Catholic religion, but it was put down, and hundreds of those who had participated in it were hanged.

A prelude to the Great Rebellion and Civil War took place in the East Riding. Charles I, anxious to gain possession of the magazines and the town of Hull, which was one of the strongest fortresses in the kingdom, appeared before the Beverley gate with a retinue of 300 persons on

April 23, 1642 and demanded admission. Sir John Hotham, the Governor, told his Majesty "that he durst not open the gates to him being intrusted by the Parliament with the safety of the town." Queen Henrietta Maria in 1642 purchased in Holland "two hundred barrels of powder and two or three thousand Armes with seven or eight Field-Pieces" which were safely landed in Keyingham Creek and transported to York where the King was. The King, with an army of about 3000 foot and 1000 horse, left York for Beverley and in July threatened to besiege Hull, unless Parliament delivered it up to him. To defeat the King's intention, Sir John Hotham ordered the sluices to be pulled up and the banks of the Humber and Hull to be cut, thereby laying under water the country for two miles on every side of Hull. Nevertheless the siege was begun but, for want of ships to bombard the town, and prevent it from being provisioned by sea, was ineffectual.

Under convoy of seven Dutch ships of war, commanded by Admiral Van Tromp, Queen Henrietta Maria returned from Holland and arrived at Bridlington Bay on February 20, 1643. She brought with her 30 brass and 2 iron cannons, with small arms for 10,000 men. She was successful in landing on February 24. Batten, the Parliamentary Vice-Admiral, cannonaded the house in which she was lodging and she had to take refuge in a ditch, but after about nine days in Bridlington, was safely conducted to York, to which place the military stores were also sent.

Sir John Hotham entered into a treacherous plot to surrender Hull to Charles, but his design was discovered and he was arrested in Beverley on June 29th, 1643, and beheaded on January 2nd, 1645.

On September 2nd, 1643, the royalist Marquis of Newcastle appeared before Hull, with about 4000 horse and 12,000 foot, and began the second siege of the town. The Governor, Lord Fairfax, on September 14th, ordered the banks of the Humber to be cut, laying the neighbourhood under water. The besiegers used red-hot shot but in spite of all their efforts had to abandon the siege on October 11th.

On July 2nd, 1644, the great battle of Marston Moor, near York, was fought and resulted in a decisive victory for the Parliamentary forces, 3000 or 4000 Royalists being slain (as compared with 1500 Parliamentarians) and 1500 Royalists being taken prisoners. York surrendered to the Parliamentary forces on July 16th and, when Pontefract Castle was finally taken in 1649, the Civil War in Yorkshire came to an end.

On September 23rd, 1779, Captain Pearson in the *Serapis* of 44 guns, with the *Countess of Scarborough* of 22 guns, was convoying the Baltic fleet, when he was attacked, off Flamborough Head, by the Anglo-American Paul Jones in the *Bonhomme Richard* of 40 guns, with the *Alliance* of 36 guns and the *Pallas* of 32 guns. After a sanguinary battle Pearson had to haul down his flag, but his convoy escaped and the *Bonhomme Richard* sank.

18. Antiquities—(*a*) Prehistoric.

As mentioned in Chapter 11, Palæolithic antiquities are practically absent from the East Riding. The barrows (burial mounds) of the Wolds have afforded much information as to the Prehistoric inhabitants. They are of two chief types—Long Barrows, which contain the remains of long-headed (dolichocephalic) Neolithic man and Round Barrows, which are usually attributed to the Bronze Age. It is to the late Canon W. Greenwell (collection in British Museum) and the late Mr J. R. Mortimer (collection in Hull Museum) that we are chiefly indebted for our knowledge of East Riding barrows. Long Barrows are few in number. Eight have been opened at, respectively, Rudston, Kilham, Willerby Wold, Heslerton Wold, Helperthorp, Westow, Gate Howe (Hanging Grimston) and Market Weighton. They vary in length from 75 to 255 feet and in breadth from 35 to 75 feet; in height at present from 1 to 7 feet.

Owing to the fact that stone implements of Neolithic type continued to be used during the Bronze Age, it is often difficult to say to which age an implement belongs, unless we adopt the somewhat arbitrary expedient of classifying implements found loosely on the ground as Neolithic, whilst those found in the Round Barrows are regarded as of the Bronze Age. Neolithic stone (including flint) axes or celts, chisels, perforated stone axes, axe-hammers, hammers, small stones for grinding, whetstones, flint flakes, saws, scrapers, borers, knives, arrow-heads (leaf-shaped, triangular, or barbed), flaking tools or fabricators, and sling stones occur chiefly on the Wolds and may be seen in the museums in London, York, and Hull.

Neolithic flint arrow head with parallel flaking, Bridlington. (*Brit. Museum*)

At the N.E. corner of the churchyard at Rudston there stands a celebrated megalithic monolith of Jurassic "moor grit" (compare Devil's arrows at Boroughbridge, West Riding). The part above ground measures 25 feet 4 inches in height, by 5 feet 9 inches to 6 feet 1 inch in width, by 2 feet 3 inches to 2 feet 9 inches in thickness. It is probably sepulchral and of Neolithic age.

In 1880 Mr T. Boynton discovered at West Furze and later at Round Hill, both near Ulrome, and at Barmston, Gransmoor, and Little Kelk, remains of lake dwellings. At West Furze there were two successive platforms consisting of trunks of trees placed horizontally and held in position by stakes. Stone implements in the lower platform and a bronze spear-head between the platforms indicate a possibly Neolithic to Bronze age.

The Round Barrows of the Wolds are circular and either conical or inverted bowl-shaped. They range from 15 to 125 feet in diameter and from 1 to 22 feet in height. Unburnt bodies are more than twice as numerous as burnt bodies in the Round Barrows.

All the types of flint and stone implements enumerated under the heading of Neolithic occur also in the Round Barrows and appear to have been specially made for the funeral. The only bronze implements not ornaments found in the Round Barrows are the plain unflanged flat celt (axe) [at Butterwick only], dagger, knife-dagger, knife, drill, and awl. This indicates that the Round Barrows belong to the *Early* Bronze Age. Unconnected with barrows there have been found *inter alia* in the riding winged celts or palstaves and a bronze mould for casting them at Hotham Carr ; socketed celts, chisels, gouges, ring, etc., from a hoard at Westow, a halberd at Bridlington, socketed spear-heads at Swine and Hutton Cranswick, and many other articles of Bronze age.

Pottery of the Round Barrows.

The Cinerary Urns range in height from 8 inches to about 3 feet and about the same in maximum width.

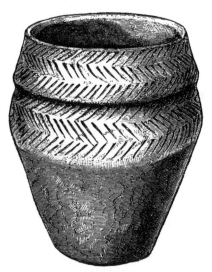

Cinerary Urn, Goodmanham ½. (*Brit. Museum*)

Their usual form is that of a long truncated cone below, its base in contact with that of a short truncated cone above. "Incense cups" vary from over an inch to 4 inches in diameter and from 1 to 3 inches in height. They are usually perforated and may have been chafers for conveying fire. Food-vessels are from 3 to 8 inches in height and of varied form. Drinking Cups (Beakers) are from 5 to 10 inches in height by from $3\frac{3}{4}$ to $6\frac{1}{4}$ inches in diameter and of two principal forms—(1) elongated tulip shaped, (2) thistle shaped ; they may also have been used for containing food for the dead. All this earthenware is ornamented with lines, triangles, chevrons, dots, etc.

The prehistoric entrenchments of the Wolds are very remarkable, on account of the enormous amount of labour their construction must have involved. The best known is the so-called Danes' Dyke, a rampart about 18 feet high, with a ditch 60 feet wide on its western side, which runs from north to south, for $2\frac{1}{2}$ miles, across Flamborough Head. General Pitt-Rivers cut a section across it in 1879 and found flint chips on the interior slope, indicating that the work cannot be more recent than the Bronze Age.

In an area of 75 square miles between Acklam, Huggate, Cowlam, and Garton-on-the-Wolds Mr Mortimer mapped 80 miles of entrenchments, often double and triple. As they sometimes cut through, or deviate to avoid, Round Barrows they must be newer and, as they are themselves cut by Roman roads, they must be pre-Roman.

The Late Celtic (Early British) or Early Iron Age inhabitants of the Wolds did not cremate their dead but buried them, usually in cemeteries, as at the so-called Danes' Graves, about 200 in number, $3\frac{1}{2}$ miles north of Driffield, where unornamented food vessels, containing originally pork, sometimes accompanied the bodies. The Britons, before the Roman occupation, were famous for enamelling in colours on bronze, more particularly horse trappings (*e.g.* bridle bits in the British Museum from Rise, 9 miles N.N.E. of Hull).

Late Celtic iron swords have been found at Thorpe near Rudston, North Grimston, Bugthorpe, Grimthorpe near Pocklington (with bronze plates of a shield); the last two, with beautiful bronze scabbards, are in the British Museum. The most remarkable Late Celtic antiquities of the riding are the "chariot burials." The

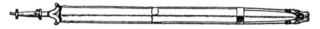

Sword and scabbard, Grimthorpe ⅑. (*Brit. Museum*)

so-called King's Barrow at Arras, east of Market Weighton, contained the skeleton of a man, the skulls of two pigs, two wheels 2 feet 11 inches in diameter, with their iron tires and bronze-coated iron fittings of the nave, the skeletons of two small horses, with their bronze-coated iron bridle bits and bronze-ended linch-pins. Some of these objects are in York Museum. Other chariot burials occurred at Westwood (Beverley), No. 13 of the Danes' Graves near Driffield, near Hunmanby station, etc.

19. Antiquities—(*b*) Roman.

The importance of fortresses and good roads for keeping in subjection the peoples they conquered was well known to the Romans. The road afterwards called Ermine Street led north from London to Lindum (Lincoln) and thence, in a remarkably straight line, to Winteringham, on the south side of the Humber, whence there was a ferry to Brough (possibly the Petuaria of

Ptolemy). At Brough there are traces of a camp and there have been found Roman coins, pottery, bronze writing styles, an iron spear-head and a piece of lead with the inscription BREXARC which puzzled antiquaries; it should be read BR · EX · ARG[ENTO] = " British (lead) de-silverised." Warburton's map of Yorkshire (1720) shows a Roman road, diverging north-eastward, from Brough to Rowley and beyond, but no traces of it are now visible. The main road ran northwards from Brough to South Cave, where in 1890 a pig of lead 22 inches × 5 inches, weight 135 lbs., was found. It is inscribed

C · IVL · PROTI · BRIT · LUT · EX · ARG

= " Gaius Julius Protus British [lead from] Lutudarum [Wirksworth, Derbyshire, mines] desilverised." The road continued northward, by South Newbald, Sancton, Londesborough Park, and Warter (Roman coins and ornaments) to Millington, ¾ mile north-east by north of which remains of a circular and three rectangular build-ings, tesselated (mosaic) pavements, and Roman coins were found in 1745. From Millington the road continued by Garrowby Top and Leavening Wold to Malton, where there is a Roman camp. Half a mile south of South Newbald the road bifurcated. The western branch ran north-westward by Thorp-le-Street and Barmby Moor Common to Stamford Bridge (Derventio). Warburton's map showed it crossing the Derwent at Kexby, for Dunnington and York (Eburacum).

Eburacum became an imperial residence, the seat of justice, and the headquarters first of the 9th, and for nearly 300 years, of the 6th Legion. It was one of the chief, if not the chief Roman city in Britain. The Roman fortress was on the left bank of the Ouse. It formed a rectangle, measuring 540 yards from N.E. to S.W. by 470 yards from N.W. to S.E., enclosing over 52¼ acres, and surrounded by a wall 4 to 5 feet thick. At each of the four angles there was probably a tower, besides many minor towers in the walls. The great tower at the N.W. angle still exists in the grounds of the Yorkshire Philosophical Society as the "Multangular Tower," crowned by a medieval superstructure. It has ten sides. The fortress had four principal gates, one on each side, and the remains of one of them has been discovered beneath Bootham Bar. Numerous remains of Roman tesselated pavements, inscriptions, altars, sarcophagi, sculpture, pottery, glass, fibulæ, bracelets, finger-rings, pins, coins, and other objects have been found, and most of them are to be seen in the museum of the Yorkshire Philosophical Society.

From York a road runs, chiefly along a glacial moraine, to Stamford Bridge and onwards to Fridaythorpe. The late Mr J. R. Boyle held it to be a pre-Roman British road, because (he writes) it does not connect Roman camps and does not run in straight lines. It is usually considered Roman and the Romans may have adopted it. Warburton's map shows a Roman road (of which no trace remains) branching from that just mentioned on the west side of the Derwent and running northwards, to continue as Wade's Causeway in the North Riding. On the east side of the Derwent a Roman road is supposed to have branched northward by Gally Gap to Malton, joining that from Brough first described, the roads running on

each side of the Derwent being connected by a road
crossing the Derwent near Firby, of which there is no trace.
To return to the York Fridaythorpe road. After crossing
the Derwent it runs to Garrowby Top, where it is crossed
by the road from Brough. About two miles east of Gar-

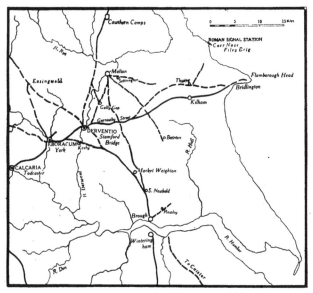

Map of Roman Roads of the East Riding

rowby Top the road bifurcates. The southern branch
runs E.N.E. about ½ mile to the south of Fridaythorpe
and is crossed by the Malton and Driffield railway 1 mile
S.E. of Sledmere and Fimber station. At this point it is
crossed by a Roman road from Malton and a Romano-
British cemetery occurs. The road continues to the Sled-

mere Monument (to Sir Tatton Sykes) and its course becomes doubtful. Some believe it to have continued by Danes' Graves, Kilham, and the road called Wold Gate to Bridlington and Flamborough Head ; others think it passed a mile north of Kilham by Rudston to Bridlington. The other (northern) branch, which diverged west of Fridaythorpe, runs by Fimber, Sledmere, Cowlam, Octon (where there is a rectangular camp, supposed to be Roman), Thwing (where a branch diverges towards Filey) and by Rudston to Bridlington. A Roman road from Malton and its bridgehead, Norton (where there was a camp), by Settrington Brow, Wharram-le-Street, Sledmere, and Fimber station crosses the southern road to Bridlington at the Romano-British cemetery and runs south-east, passing west of Wetwang, 1 mile west of Tibthorpe, and west of Bainton. Another runs from Malton by Settrington High Street towards Weaverthorpe.

In 1857 a Roman signal or coastguard station was discovered at Carr Naze, Filey, consisting of an oblong walled courtyard, containing a room 15 × 12 feet and five socketed stone pedestals (now in the Crescent Gardens, Filey), arranged quincuncially (at the angles and centre of a square), with a sixth close to the north-east one. They must have supported a wooden structure and were accompanied by Roman pottery, querns, iron spear- and arrow-heads, etc. About 1839 a Roman tesselated pavement, measuring about 4 yards by 3 yards, was discovered, adjoining the road from Rudston to Kilham. In 1904 remains of three Roman tesselated pavements were found $1\frac{1}{4}$ miles north of Harpham, near Burton Agnes. The best one, 16 feet × 17 feet, representing a square maze,

with all its angles right angles, is in the Municipal
Museum at Hull. They were accompanied by pottery,
beads, stone roofing-tiles, painted wall plaster, coins
(ranging from 253 to 273 A.D.) of Gallienus, Victorinus
and Tetricus, and traces of a hypocaust (underground hot
air chamber and flues). In 1875 a Roman kiln (or hypocaust ?) was discovered 1 mile south-west of Langtoft and
in 1874 a similar structure was met with at Etton. About
1721 two tesselated pavements were discovered in fields at
Bishop Burton. One mile east of Langton, Roman
pottery, tesseræ of a pavement, and Roman tiles were
discovered in 1863; and in 1899 on the opposite (east)
side of the road from Birdsall to Malton 33 square pillars
of a Roman hypocaust, Roman tiles, pottery, and a coin
of Trajan (A.D. 98–117) were found.

20. Antiquities—(c) Anglo-Saxon.

Pagan Anglo-Saxon interments are usually accompanied
by relics and are therefore more interesting than the
graves of Anglo-Saxon Christians. The chief types of
Anglo-Saxon grave articles found in the E. Riding are :
iron objects, including swords, spear-heads, arrow-heads,
shield bosses of circular wooden shields, knives, knife-
sharpeners, bridle bits, bucket hoops, buckles, bodkins,
shears, and girdle-hangers (chatelaines); bronze objects,
including strap ends, buckles, clasps, thread boxes, brooches
(square-headed, annular, and penannular), armlets, pins,
tweezers, gold pendants and bracelets, silver ear-rings, and
annular brooches ; necklaces of varied materials, bone
combs and pins, bone, chalk or earthenware spindle

whorls, earthenware cinerary urns and food vases. The chief localities are Garton Slack (near Garton-on-the-Wolds), Kelleythorpe Barrow and Cheesecake Hill, both near Driffield, Hornsea Hydropathic, Sancton, and Uncleby ($\frac{1}{2}$ mile east of Kirby Underdale).

The numerous cinerary urns (in Hull Museum) found at Sancton are usually globular in shape, about 10 inches high and often decorated with circular or oval bosses formed by pressure from within and with lines, circles, crosses, squares, etc. The food vases were globular, hemispherical, or shouldered and without ornament.

In the Hull Museum, the "Roos Carr Images" consist of three small wooden images 13 to 16 inches long of men standing in a serpent-shaped boat, and two more from another boat. They were found in 1836, near Roos, and are believed to be of Scandinavian pre-Viking age. Anglian and Anglo-Danish sculpture (including sundials) of pre-Conquest date is much less abundant in the East than in the West Riding. It is chiefly in the form of stone crosses and their shafts, usually adorned with interlaced or plaited patterns, figures of birds, beasts, dragons or snakes, and finally human figures. It is recorded from about 15 sites, sometimes inside (i) or built into the walls outside (o) the church or at the vicarage (v). In York there are 32 stones with carving of a pre-Norman character, of which eight have come from other places. Twenty-five of these stones are in the Hospitium of the York Museum. Among the most noteworthy East Riding fragments there is a part of a cross shaft with beasts and human figures from Folkton. A shaft fragment with dragons remains at Folkton. At Londesborough a cross head, with unique interlaced design, and a dial, supposed to be Anglian, are built into the tympanum of the Norman south door of the church. At North Frodingham there is a wheeled cross head (v) ; at Nunburnholme a Danish eleventh century cross shaft (in the churchyard) with the Madonna, saints, abbess, etc. ; at Sherburn 10 pre-Conquest fragments (o) ; at Barmston a hog back (recumbent monument) (v) ; at

Pre-Conquest Cross, North Frodingham

Filey, on the tower stair, an interlaced stone; Hunmanby, crosshead (o) Leven, shaft fragment (v) Lissett, bear's head (o) Little Driffield, fragments (i and o). At Aldbrough (i) is a circular sundial (from an earlier church), inscribed (translation) "Ulf had this church built for himself and for the soul of Gunnvőr." At Weaverthorpe, built in over the S. door, is a dial, with Latin inscription "IN HONORE S(AN)C(T)E ANDREÆ APOSTOLI HEREBERTUS WINTONIE HOC

MONASTERIUM FECIT IN TEMPORE Rᚻ." Mr W. G.
Collingwood, the authority on these sculptures, states "the
connection of this inscription with Archbishop Osketel
and King Regnald is illusory." At Kirkburn three un-
inscribed sundials (o) may be Anglian.

21. Architecture—(a) Ecclesiastical.

Before describing the churches, religious houses and other
important buildings of the East Riding, a few words may
be devoted to the various styles of English architecture.

The features most characteristic of Pre-Norman, or
Saxon, churches are absence of buttresses, presence of
pilaster-strips, quoins of very large stones or "long-and-
short work" (upright square pillars, alternating with flat
slabs) at the angles, a nave of great *proportional* length
and height with a roof of high pitch, a very plain and
relatively slender western tower, belfry windows of two
round-headed lights, with a "through" stone and "mid-
wall shaft."

The Norman Conquest started a widespread building
of churches and castles in the Romanesque style, which in
England is usually called "Norman." It is characterised
by walls of great thickness, semi-circular arches and vaults,
round-headed windows and doorways (the latter often recessed), wide, flat buttresses, and chevron, billet, nail-head,
beak-head, etc. ornaments.

Between 1145 and 1190 Transitional Norman, in which
pointed are mixed with round arches, developed into
Gothic architecture, in which, in combination with pointed
arches, there was perfected the science of vaulting, by

which the weight and thrust of arches are brought upon piers and buttresses.

From 1190 to 1245 the first English Gothic, called "Early English" is characterised by slender piers, often with clustered marble shafts, deep buttresses, lofty pointed vaults, narrow lancet-headed windows, and dog-tooth ornament.

After 1245 the windows became broader, divided by mullions into several lights, and adorned with geometrically designed bar-tracery ("Geometrical" or "Early Decorated"). From about 1315 the structure of stone buildings began to be overlaid with ornament, especially "ball flowers," the window tracery showed rich double curvature and ogee arches, the vault ribs formed intricate patterns, the pinnacles and spires were loaded with crocket and finial. This latter style is known as "Curvilinear" or "Late Decorated" and lasted until 1360.

With curious uniformity and rapidity the style called "Perpendicular"—which is unknown abroad—developed after 1360 in all parts of England and lasted until 1550, the style after 1485 being often termed "Tudor." As its name implies, it is characterised by the vertical-lined arrangement of the tracery and panels on walls and in windows, very thin shafts with octagon capitals and bases, broad four-centred arches, with rectangular hood-moulds, by the elaborate vault-traceries (especially fan-vaulting) and by the use of flat roofs and towers without spires.

The medieval styles in England ended as a result of the dissolution of the monasteries (1536–1540), for the Reformation checked the building of churches. There succeeded the building of manor houses, with Tudor windows, mullioned and transomed but square-headed. In the Elizabethan, and still more in the Jacobean styles, classic columns and ornaments were introduced, under the influence of the Renaissance, and were accompanied by strap-work ornament and pierced crestings.

In the description of the chief churches, followed by that of the Religious Houses of the riding, the following abbreviations are used : Sax. = Saxon, Norm. = Norman, Trans. = Transitional Norman, E.E. = Early English, Dec. = Decorated, Perp. = Perpendicular. An asterisk * indicates pre-Conquest sculpture, mentioned in Chapter 20. Additional churches are mentioned in Chapter 27.

Of pre-Norman work in our riding we have the two lower stages of the W. tower of Skipwith church, the S. nave door of which has beautiful twelfth century ironwork. Wharram-le-Street church has a Sax. W. tower. In York, St Mary Bishophill Junior has a Sax. W. tower (possibly rebuilt) and recessed tower arch.

The church of Weaverthorpe* has a W. tower, almost Sax. in appearance, an aisleless Norm. nave, chancel arch, and S. doorway, and a Norm. font, with a diaper of circles and octagons, like that at Rudston.

North Newbald church has an aisleless Norm. nave, central tower (lower part) and transepts, a Perp. chancel, and E.E. font. Kirkburn* church has an aisleless Norm. nave, W. tower (lower part), chancel arch, S. doorway and font. Garton-on-the-Wolds church has an aisleless Norm. nave and W. tower, modern S. doorway and chancel. Etton church has a broad Norm. W. tower, a very fine E. tower arch, and a Trans. or E.E. south aisle. Wharram Percy has a Norm. W. tower. Goodmanham

Norman Chancel Arch, Goodmanham Church

has Norm. chancel arch, S. doorway and W. tower; the
nave arcade (N.) and chancel are E.E. Of the fonts, the
older is Norm. the newer Perp. The interesting church
of Stillingfleet has a Norm. N. and a magnificent Norm.
S. doorway. The ironwork of S. door is of about 1145
and shows a viking's ship, etc.; the N. arcade of nave and
chancel and base of tower are E.E.; the Moreby (S.)

Weaverthorpe Church

chapel is Dec. Rudston has a Norm. W. tower and font,
rest of church Dec.

Bishop Wilton has a Norm. chancel arch and S. doorway, an E.E. nave and a Dec. or Perp. W. tower. Flamborough has a Norm. chancel arch and font, a magnificent
Perp. rood screen, and a brass inscribed to Sir Marmaduke
Constable (d. 1518). Kilham has a Norman S. doorway
recessed in six orders. Both St Denis and St Margaret's,

Rood Screen, Flamborough Church

Walmgate, York, have beautiful Norman doorways. Cottam and Cowlam have Norman fonts of great interest, that of Hutton Cranswick is in the York Museum (Hos-

Norman Font from Hutton Cranswick, now in York

pitium). Filey* nave illustrates the passage from Trans. to E.E. ; the central tower, chancel and transepts are E.E. Great Driffield has an E.E. font and round-headed E.E. nave arcade and N. and S. doors, with Perp. W. tower. In

the beautiful church of Hedon the transepts and chancel
are E.E., the nave Dec., the central tower and handsome
font Perp. Pocklington has E.E. nave, S. doorway and
N. transept, with eastern aisle, Perp. chancel, W. tower
and square Norm. font. Langtoft has E.E. West tower
and S. arcade of nave, and a Dec. chancel.

Patrington church is very beautiful and almost entirely
Dec. It has a nave with N. and S. aisles, transepts with E.

Filey Church from S.W.

and W. aisles, aisleless chancel, central tower, with
octagonal corona, a lofty spire, and a vaulted parvise (priest's
chamber) over S. porch. In the E. aisle of S. transept is
the recessed Lady Chapel and on N. side of chancel the
Easter Sepulchre, in four compartments, with sleeping
soldiers below, and above, the Saviour rising from the
tomb. The beautiful church of St Mary, Beverley, has
remains of Late Norman work in the S. door of nave and

of E.E. work there and in the responds at the E. end of
the nave aisles. The rest of nave aisles, S. arcade and S.
aisle of chancel, N. and E. wall of vestry or chapel (on E.

Patrington Church from S.E.

of N. transept) and E. wall of the S. transept are Geometrical ; the N. arcade and aisle of chancel and sacristy
N. of it Curvilinear ; the west front, nave arcade, S. porch,
main portion of transepts, central tower, all the clerestories

and the font are Perp. The sixth pier from the W. in the
N. arcade of nave has five figures of minstrels, whose guild

Easter Sepulchre, Patrington Church

paid for it after the collapse of the tower in 1520. Holy
Trinity Church, Hull, has late Dec. transepts, Broadley

Chapel and chancel, all built of fourteenth century brick externally, Perp. nave, aisles, and central tower. The large sixteen-sided Perp. font resembles that at Hedon. In the S. aisle of the chancel is an altar tomb, with alabaster effigies, attributed to Sir William de la Pole (d. 1366) and his wife. He was a merchant of Ravenserodd and afterwards of Hull, of which he was the first mayor. Bainton and Cottingham have beautiful Dec. (Curvilinear) churches, the latter with Perp. chancel and central tower.

Many examples of Perpendicular work have been mentioned. The chapel of South Skirlaugh is a good example of the style. St Mary's, Lowgate, Hull, consists of six Perp. bays, of which the three to the E. constitute the chancel. There are a N. and two S. aisles, of which the southern dates from 1860–3. The original W. tower was replaced by a new one in 1696, since cased and raised by Mr (Sir) G. G. Scott, in 1860–3. Paull has a Perp. church with central tower. Sancton has an octagonal Perp. W. tower.

Before describing the Religious Houses of the East Riding, we may devote a few words to the monastic orders to which they belonged. The Benedictines, or Black Monks, lived according to the rule of St Benedict, who was born at Nursia (Norcia) in Italy, about A.D. 480. The name of the Cistercians (Grey or White Monks) is derived from the abbey of Cîteaux (*Cistercium*), founded near Dijon in 1098. The Cistercian was the chief offshoot from the Benedictine Order. Monks were not necessarily priests, but regular canons were essentially priests. The Augustinian Canons (Austin Canons) followed the so-called rule of St Augustine, drawn up towards the end of the eleventh century. They resembled monks in living in communities, but served as priests of the churches under their patronage.

The Gilbertine Order was founded by St Gilbert of Sempringham (in Lincolnshire), born about 1083–1089. It was a double order, men and women separated but living side by side. Unlike monks and canons regular, friars were not necessarily attached to a religious house in one definite locality, but belonged to a wider province or order. There were four great orders of mendicant friars, Franciscans, Dominicans, Carmelites, and Augustinians.

There were two houses of Benedictine monks in York, viz. All Saints' Priory, Fishergate, of which nothing remains, and St Mary's Abbey. William Rufus laid the first stone of the original St Mary's Abbey, which terminated eastward in seven apses, of which the foundations only remain. It was destroyed by fire in 1137, and the church, of which the picturesque ruins still exist, was begun in 1271. It was cruciform, the nave of eight and choir of nine bays, of almost equal length; the transept had an eastern aisle, and there was a central tower. The chief existing remains are the wall of the N. nave aisle, with eight Geometrical windows. Between each pair of windows are blind lancets and the springers of the ribs of the vaulted roof. Beneath the windows is a Geometrical blind arcade. Part of the W. end of the nave, the arch at the E. end of the N. aisle, and part of the N. transept still remain, as well as the abbey gate-house and hospitium*, and the foundations of most of the original buildings.

The Benedictine nuns had priories at Nunburnholme, Nunkecling, Thicket, and Wilberfoss, of which nothing remains, and St Clement's Priory, Clementhorpe, York, of which a fragment of wall remains. The Cistercian monks had Meaux Abbey, of which nothing remains above ground. The Cistercian nuns had Swine Priory. The chancel of the priory chapel survives in the existing parish church, which has a Trans. nave arcade and an interesting series

Monuments in Hilton Chapel, Swine Church

of monumental effigies in the Hilton Chapel. The Carthusian monks had a priory in Hull, which has been "razed to the very foundations." The Austin Canons had Haltemprice Priory 1¼ miles S. of Cottingham, of which nothing remains ; Warter Priory, which was partially excavated in 1899 (but buried again); besides foundations an incised grave slab of the prior Thomas Bridlington (d. 1498) was found; Kirkham Priory (abbey) of which the richly ornamented Dec. gatehouse, a beautiful E.E. lancet of the church and, in the cloisters, a Norm. doorway and the fine Geometrical lavatorium, are the chief remains; Bridlington Priory church, a beautiful building of which the

Gatehouse and base of Cross, Kirkham Priory

doorway of the N.W. tower, exterior of N. side of nave, N. arcade and seven eastern bays of S. arcade of combined nave and chancel are E.E.; part of the N.W. tower and exterior of S. side of nave, the N. triforium and clerestory and part of the S. clerestory are Dec.; the S.W. tower, centre of W. front, three W. piers of S. arcade of nave, three E. windows of S. clerestory and stone screen at the base of the S. clerestory windows are Perp. Five arches of the Trans. cloister arcade are preserved in the N. nave aisle.

The Gilbertines had St Andrew's Priory, York (demolished); Ellerton Priory, of the chapel of which some

Oriel Window, Watton Priory

relics are preserved in the modern church; and Watton Priory, excavated between 1893 and 1898 and reburied, but the Prior's Lodging, with its beautiful two-storied Perp. oriel window, survives. There was a Dominican Friary at Beverley, of which part of the red brick boundary wall, etc. remains, E. of the Minster. There were medieval hospitals, which helped the poor and infirm, as well as pilgrims, lunatics and lepers at Beverley, Braceford (near Burton Agnes), Bridlington, Fangfoss, Flixton, Hedon, Hessle, Hull, Killingwoldgraves, and York. That of St Leonard at York was the largest and wealthiest. Its ruins are near St Mary's Abbey.

Collegiate Churches are so-called from having a college or chapter, consisting of a dean or provost and canons, attached to them (in most cases no longer). In York there were four—the Bedern (off Goodramgate), a college of Vicars Choral; St Sepulchre's N. side of Minster (taken down in 1816); St William's College (for parsons and chantry priests), which has a Perp. entrance doorway but is chiefly Jacobean within; and lastly the Cathedral of St Peter.

The splendid cathedral church of York has an interior length of 486 feet, and its nave, 99 feet high, consists of eight bays, the choir of nine. The central and two western towers are of almost equal height (198 and 196 feet). The building is mainly of three periods and styles of architecture. The main (western) transepts which have aisles on both sides and the largest triforium in England, are Early English. That to the south contains the most used portal of the cathedral and its façade is much loaded with ornament; that to the north has a group of lancet windows, 50 feet in height, celebrated at the "Five Sisters." The nave, the lower part of the west front, the beautiful octagonal chapter-house, and the vestibule connecting it with the N. transept are Dec. (Geometrical). The west window and the tower windows on either side of it are Dec. (Curvilinear). The East transept, the choir, with its great East window, 78 by 32 feet, the central tower, and the upper parts of the western towers are Perp. In the

York Minster, West Front

crypt are circular Norm. columns and piers, one of them with the lattice-work pattern on its shaft, which occurs in

York Minster, the Choir

Beverley Minster, West Front

Durham Cathedral and Selby Abbey church. The greatest glory of the minster, however, is its magnificent stained glass. Beverley Minster is by many considered to be the

Percy Tomb, with Early English arches and triforium
in background, Beverley Minster

most beautiful building in Yorkshire. The dominant features of its exterior are the lofty E.E. Choir, great (western) transept with E. and W. aisles and E. transept with E. aisle, the transept fronts with tiers of lancet windows, crowned in the W. transept by a great rose window and (on S.) a vesica and flanked in both transepts by octagonal turret buttresses ; the nave, chiefly Dec. ;

Howden Church. Ruined chancel on the right,
Chapter-house on the left

the two graceful Perp. W. towers 162 feet in height and the Perp. N. porch and W. part of N. aisle of the nave. In the interior the E.E. main arcades of the choir, retro-choir and transepts have clustered shafts, the blind tri-forium has a double arcade and, like the clerestory, Purbeck marble shafts. The nave arcade, triforium and clerestory, chiefly Dec., are assimilated to the E.E. portions of the building. Among many beautiful features, there is in the N. aisle of the choir, an E.E. double staircase, under a trefoiled arcade, and (adjoining the altar screen), the magnificent Dec. Percy Tomb of Eleanor, wife of Lord Percy of Alnwick, who died 1328. Of great historic interest is the stone Frith Stool, or sanctuary chair, in the N. aisle of the choir.

The very fine church of Howden is chiefly Dec. Its ruined chancel is more ornate than the nave. The central tower is Perp. and so is the exquisite ruined Chapter House. In Hemingbrough church the two E. bays of the nave arcade are Trans., the rest of the building Dec. and Perp.; the central spire is 120 feet high. The churches of Lowthorpe and Sutton-in-Holderness were formerly collegiate.

22. Architecture—(*b*) Military.

Remains of ancient military architecture in the riding are almost confined to the ruins of Wressell Castle, but it may be of interest to give a short account of fortified buildings which formerly existed.

A castle existed at Aldbrough in 1115. At Aughton, on the N. side of the church, is a motte (mound) surrounded by a ditch and on the E. side of the church is the bailey or basecourt of the castle or residence of the Aske family. Beverley had formerly five gates or bars—South bar, Norwood bar, Keldgate bar, Newbegin bar, and North bar, the only one remaining. It was built of brick in 1410 at a cost of £96. 17*s*. 4½*d*. The groove for the portcullis remains. In 1388 the prior and convent of Bridlington obtained a license to crenellate (fortify) their priory. Of these fortifications the Bayle Gate is all that remains. On Mount Ferrant, 1 mile S.E. by S. of Bury-thorpe, there was a Norman castle of wood, which was destroyed about 1150. At Cottingham was Baynard's Castle, built by William de Stuteville, by license from King John, in 1200. Leland, writing between 1535 and

The Bayle Gate, Bridlington Priory

and 1543, says: "I saw where Stutevilles Castelle, dobill dikid and motid, stoode, of the which nothing now re-maynith" [except, at present, the earthworks]. The Moat (or Moot) Hill, Driffield, was the site of a castle. Marma-duke Constable obtained a license to crenellate his house at Flamborough, probably the so-called "Danish Tower" in 1352, and licenses to crenellate a house at Sutton on

Derwent in 1292–3 and at Harswell in 1302–3 were also granted. At Castle Hill, Hunmanby, are earthworks, attributed by Leland to a castle of the earl of Northumberland. In 1308 Henry Percy obtained a license to fortify his castle at Leconfield. The great moat is all that remains of this Earl of Northumberland's castle (demolished in the reign of James I), of which Leland writes—"Leckingfeld is a large house, and stondith withyn a great mote yn one very spatius courte. 3. partes of the house, saving the meane gate that is made of brike, is al of tymbre. The 4. parte is fair made of stone and sum brike." At Lockington, near Hall Garth, is a motte called Coney Hill or Moat Hill, surrounded by a ditch and with traces of a bailey. At Skipsea, the Castle Hill was the site of a castle, erected by Drogo de Bevrère, who, in 1086, was the first lord of the Seigniory of Holderness. It was the capital seat of his barony and was, by royal mandate, ordered to be destroyed in 1221. The motte was in Skipsea Mere and there was a huge bailey. One mile S.W. by S. of Swine church is Castle Hill, the undoubted site of a castle (Branceholme ?) in 1353. The castle of Wheldrake was destroyed by the citizens of York, by leave of Stephen, in 1149. Of Wressell Castle Leland writes: "Most part of the basse courte [bailey or basecourt] of the castelle of Wresehil is al of tymbre. The castelle it self is motid about on 3. partes. The 4 parte is dry where the entre is ynto the castelle.... In the castelle be only 5. towers, one at eche corner almost of like biggenes. The gate house is the 5." The castle, as built by Thomas Percy about 1380–90, consisted of buildings surrounding a quadrangular court. After the Civil War, Parliament insisted on Wressell Castle and other fortified mansions being made untenable and three sides of the castle were demolished in 1650. The chief remains are the two towers of the south side and the building connecting them.

The ancient defences of Hull were sanctioned by Edward II in 1322. They consisted of a moat and wall, with towers and gatehouses, built of brick. The site of their north and west portions is now occupied by the Queen's Dock, Prince's Dock, and Humber Dock and

Wressell Castle

that of the south wall by the south side of Humber Street. The gates were the North Gate, Low Gate, Beverley Gate, Myton Gate, Hessle Gate, and the Tower Gate leading to the jetty on the Humber. These walls were 2610 yards in length or 30 yards less than 1½ mile. By order of Henry VIII, in 1541, a castle with walls 19 feet thick, and 142 feet square (with E. and W. projections), and a N. and S. blockhouse, each in the form of a club on a playing card, were built of brick on the E. side of the River Hull

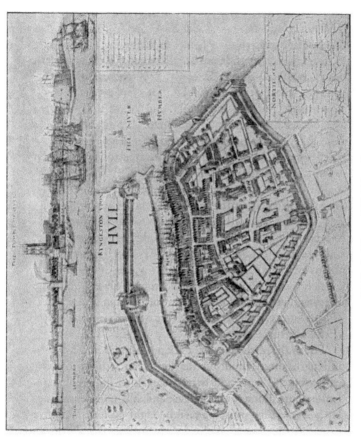

Hollar's View of Hull in 1640

and connected with one another by a wall about 1200 yards long. The Dutch engraver Hollar's bird's eye view of Hull in 1640 shows the walls on the W. bank of the River Hull to have had 33 towers, including 6 round and 6 gate towers. In 1681 a triangular citadel, with an eastern moat, was built on the E. bank of the River Hull, incorporating the castle and S. blockhouse. When the docks were made, the walls on the N. and W. of the city were demolished, between 1775 and 1826. The N. blockhouse was taken down in 1801–2 and the demolition of the citadel began in 1863. One of its watch towers has been rebuilt in East Park.

William the Conqueror built two castles at York. The first was situated on the tongue of land between the Ouse and the Foss, where the existing castle stands ; the second was on Baile Hill on the opposite (right) bank of the Ouse, and was completed in eight days. The chief existing remains of the ancient castle are 120 yards of the original curtain wall, two three-quarter round towers, and the keep, known as Clifford's tower, of quatrefoil plan, built 1245 to 1259, when the wooden castle was replaced by a stone structure.

The walls of the city of York are 2 miles and 15 yards in length, mostly built on the top of an artificial bank of earthwork. They are of white magnesian limestone, have battlements, numerous bastions, and still more numerous buttresses. They were pierced by eight posterns, and five great gates or "bars," the older part of which is Norman. These bars—Bootham on the N.W., Monksbar on the N.E., Walmgate on the S.E., Fishergate on the S., and Micklegate on the S.W.—had originally no portcullises, but in the Decorated period these were added and still remain in Bootham, Monk, and Walmgate Bars. The gatehouse had also added to it a superstructure with turrets or bartizans. In front of each gatehouse was a rectangular barbican or enclosure, with battlements and an outer gate. This has been restored at Walmgate.

Micklegate Bar, York

23. Architecture—(c) Domestic.

William Harrison, in his *Description of Britain*, 1577, writes—"The greatest parte of our buylding in the cities and good townes of Englande consistith onely of timber." Away from wooded districts smaller quantities of timber were used "whereunto they fasten their splintes or radles,

and then cast it all over wyth clay to keepe out the winde."
Leland, writing 1535–1543, says—"The toune of Beverle
is large and welle buildid of wood."

Very few timber or half-timber houses still exist in the
East Riding. The King's Head Inn, High Street, Hull,
which, when it was demolished in 1905, was the oldest
house in Hull, was half-timbered and had overhanging
upper stories. There still remain two houses with overhanging stories (presumably half-timbered) in High Street,
one of them at the entrance to George Yard. It is curious
that Leland writes that in Michael de la Pole's time
(1330–1389) "the waul" [of Hull] was made al of brike,
as most of the houses of the toun that tyme was." Mr
W. H. Crofts informs me that there are a half-timber
cottage at South Dalton and one or two small ancient
half-timber houses in Beverley. In York there is a timber
house at the end of the Pavement and a house in Newgate,
near the Shambles. The Merchants' Hall, Fossgate, York,
consists of a lower storey of brick, with stone dressings,
formerly used as a medieval hospital, and an upper storey
of half-timber, probably built between 1357 and 1368.

Among ancient buildings of stone or brick the Manor
House of the Bishops of Durham, at Howden, is noteworthy. There was a Manor House in Howden in 1153,
but we have no details of it. The buildings, of which
remains survive, were round the sides of a quadrangle.
All that remains are some of the walls of the quadrangle,
the Great Hall 62 feet by 36 feet internally (now divided
up), and the two-storied vaulted stone-faced porch (now
used as a dairy). It has a blocked round-arched doorway,
a Perp. square-headed window and embattled parapet and,

Half-timbered Houses, Stonegate, York

with the hall, was built by Bishop Skirlaw (1388-1406).
It is now a private house and lies east of the vicarage.
Another relic of the Manor House is an archway (west of
the vicarage), erected by Cardinal Langley (1406-1437).
Paull Holme Tower is the remains of a manor house. It
is about 30 feet high with battlements and loop-hole
windows and, from the roses carved on it, may date from
the reign of Henry VII. South Frodingham Hall, 2 miles
W.S.W. of Withernsea, is of brick with mullioned windows
and was built in the reign of Henry VIII and added to by
the Listers in that of Elizabeth. Knedlington Old Hall,
1 mile W. of Howden, belonged to Sir John Gate in the
time of Henry VIII and is a good specimen of Tudor
domestic architecture. Heslington Hall, 2 miles S.E. of
York, was built in 1578, in anticipation of a visit by
Queen Elizabeth. It is of brick, with stone dressings. It
has bay windows, three storeys in height, and a fine courtyard. The Old Hull Grammar School near Holy Trinity
Church was built in 1583 and restored in 1883. Wilberforce House in High Street, Hull, was built for the Lister
family about the end of the reign of Queen Elizabeth. It
is of brick, was altered by the Wilberforces in the
eighteenth century, was the birthplace of William Wilberforce the abolitionist, and is now a museum. The Elizabethan red brick school-house and hospital at Halsham
were founded by the will of Sir John Constable in 1579.

Perhaps the most beautiful private house in the riding
is Burton Agnes Hall, built of brick, with stone dressings,
in 1601-3, added to in 1628. It has semi-circular and
half-octagon bay-windows, in three storeys, and the handsome gatehouse is of the same Jacobean date. Howsham

Hall, at the foot of the Malton Gorge, built by Sir W.
Bamburgh in 1612 of dressed stone, has a projecting
central two-storied porch, flanked by two-storied bay-
windows and crowned by semi-circular crestings. The
King's Manor House, York, occupies part of the site of
the Abbot's House of St Mary's Abbey and is partly Elizabethan but chiefly Jacobean. There are several Jacobean

Gatehouse, Burton Agnes Hall

doorways and a Carolean one, with the arms of the great
Earl of Strafford. Other Jacobean houses are the Manor
House, Bishop Burton ; Beswick Hall (S. of Watton) ;
Arram Hall (2 miles W.S.W. of Atwick) and to some
extent the Treasurer's House at York. Burton Constable
Hall, 3½ miles E. by N. of Swine, is the capital seat of the
Seigniory of Holderness. It dates from about 1630 and
forms an imposing pile. It is built of brick, with stone

dressings, round three sides of an entrance-court and has many large mullioned and transomed windows, with bays

Jacobean Doorway, the King's Manor, York

and oriels. The fine red brick Hall of Winestead, near Patrington, formerly the mansion of the Hildyards, dates from 1710. In Hull, No. 60 High Street, is the old home of the Maister family. It has a handsome entrance-hall, lighted from an octagonal lantern in the roof.

24. Communications—Past and Present —Roads, Canals, Railways.

The earliest roads in Britain were mere trackways connecting hamlet with hamlet. The excellent system of Roman roads has been described in Chap. 19. After the departure of the Romans their roads were adopted by their successors in later times. The repair of roads and bridges was part of the *trinoda necessitas* or triple obligation of landowners of Anglo-Saxon times, but was often neglected. Other means of maintaining roads and bridges in the Middle Ages were special gilds, endowments, tolls, and the pious offerings at the chapels on the bridges. These chapels still exist at Wakefield and Rotherham, and St William's Chapel on the old Ouse Bridge at York was only demolished in 1810.

Mr J. R. Boyle has laid down rules by which roads may be allocated to one or other of seven periods, as follows :

1. Pre-Roman or British.

2. Roman.

3. Anglo-Saxon, covering the period from the early Teutonic invasions to the arrival of the Danes.

4. Danish, covering the period from the Danish invasions to the Norman Conquest.

5. Medieval, extending from the Norman Conquest (1066) to the great period of enclosure (1450–1550).

6. Recent, from the period of enclosure to the middle of the eighteenth century.

7. Modern, from the middle of the eighteenth century to the present time.

For example, the road from York to Fridaythorpe, mentioned in Chapter 19, p. 98, divides eight townships on the north from nine on the south. It must have been in existence before these 10 Anglo-Saxon and 7 Scandinavian townships and is, therefore, either Roman or pre-Roman. As it does not join Roman camps and is not straight, it is pre-Roman. We must distinguish between road termini (such as Beverley) and roadside settlements. Take the road from Hessle to Beverley. In its whole length of $8\frac{1}{2}$ miles it only passes through one village, the Danish village of Willerby, obviously a roadside settlement, hence this road is pre-Danish or Anglo-Saxon. The road from the Anglo-Saxon Barmston to the Danish Carnaby must be at least as late as the second settlement, and belongs to the fourth (Danish) period. A road not coming under the preceding categories is of post-Conquest date. Before enclosure began, about 1450, the open-field system prevailed; the open arable fields formed strips or balks, and the country was not cut up by hedgerows and fences. Take the three roads from Hull to, respectively, Beverley, Anlaby, and Bilton (which, with the road to Hessle, date from 1303). In not one of these cases does the line of a single hedgerow on one side of the road coincide with the line of a hedgerow on the other side. These roads belong to Period 5. The two modern roads from Hull to Hedon and Ferriby, respectively, intersect hedgerows and belong to Period 7. Roads which have been formed since the great period of enclosure, and before the date of the indispensable Act of Parliament authorizing the enclosure of pasture lands, belong to Period 6. Until about the middle of the seventeenth century the ordinary mode of travel was on foot or on horseback. In 1658 a stage coach ran thrice a week from London to York in four days for eleven shillings.

The Rev. Robert Banks, vicar of Hull, wrote to W. R. Thoresby of Leeds, December 29, 1707, "the ways in Holderness at this time of the year are next to impassable, and some have lost their lives who have ventured through them." The Rev. W. Dade wrote October 7, 1777, "I was however happy in being enabled yesterday to visit Aldbro' church; in a month's time it will be inaccessible on account of the road."

The great coaching era began in 1786, when the Royal Mail was first carried by coach, and ended in 1842, when the Royal Mail coach which had run regularly from London to York since 1786 finally went off the road.

The cost of transporting goods by water was so much less than that of carriage by road that rivers were early made use of, though navigable canals have only been constructed in comparatively modern times.

The conservancy of the Ouse, Aire, Wharfe, Derwent, Don, and Humber was entrusted by Charter to the municipality of York in 1462. In 1531 the city of York obtained an Act for pulling down and avoiding of fish-garths, piles, stakes, hecks, etc. in the Ouse and Humber. It was not until 1757 that the first Naburn lock on the Ouse was opened; in 1888 an additional lock capable of passing vessels of 300 to 400 tons burden was opened there.

The Ouse is navigable from the confluence of Ure and
Swale to its mouth at Trent Falls (60 miles 6 furlongs).

The Archbishop of York, Lord of the town of Beverley,
and owner of the soil on both sides of the river Hull, took
toll from boats and small vessels on that river. His privilege

A Yorkshire "keel" on Beverley Beck

at the town of Hull, in 1213, was to have a free passage
along the river's midstream, of the breadth of 24 feet.
Through the intervention of Archbishop Walter Gifford
the river was made navigable for ships in 1269, by the
removal of weirs and fishery-fences. It is now navigable

for 20 miles, from its junction with the Driffield Navigation, at Struncheon Hill lock, to its confluence with the
Humber. In Chapter 16, p. 79, the canal of Beverley
Beck has been mentioned. It is 67 chains (1474 yards) in
length and can accommodate vessels 65 feet long in its lock.

The river Derwent was made navigable up to Malton,
38½ miles, by an Act 1 Anne (1702) and, in the beginning of the nineteenth century, the navigation was extended to Yeddingham, but in 1846 the Rye and Derwent Drainage Act was obtained for removing mills,
mill-dams and locks which obstructed the drainage, and
the navigation now ends at Malton. The Driffield Canal
Act was passed in 1767 and the canal was opened in
1772. The main line extends from Driffield Wharves to
the junction with the river Hull, 7 miles, and in addition,
the Frodingham Beck branch and the branch to Corps
Landing each measures 1¾ miles. The Market Weighton
Canal Act was passed in 1772 and the route was surveyed
by the celebrated John Smeaton. The canal, which serves
both drainage and navigation, was originally a little over
9 miles long, but the portion between Sod House lock and
Weighton River Head was closed for navigation by Act
of Parliament of 1900, leaving 6 miles in use. During
and since the Great War the level of water for drainage
purposes has made navigation impossible. The portion of
old canal is now very shallow and almost dried up in places.
It is much overgrown by jungles of reeds. The Leven
Canal was authorised by Acts passed in 1800–1 and
1804–5 and was opened in 1802. It is 3 miles 26 chains
long and extends east to west from Leven to the river
Hull. The Pocklington Canal was constructed under an

Act passed in 1814. It is 9 miles 30 chains in length and extends from a mile south of Pocklington to the Derwent at East Cottingwith. The upper portion of this canal appears to be practically derelict.

From the sea to Hull the channel of the Humber is fairly constant, but from Hull to Trent Falls the bed of the river is constantly varying. At Hull spring tides rise 20 feet 9 inches, neap tides 16 feet 3 inches.

The following table shows the lengths of the canal systems and the tonnage of goods transported in various years:

	Miles	1905	1913	1920
Ouse (passing Naburn Locks)	60¾		350.143	198,752
Beverley Beck (67 chains) .		101,540	49,925	41,816
Derwent	38½	6,076	4,286	1,345
Driffield Canal and branches	10½	32,666		
Market Weighton Canal (1898: 17,366 tons) . .	6			
Leven Canal	3½	4,546		
Pocklington Canal . . .	9⅜	1,076	1,657	37

Turning to Railways, all the railways of the East Riding belong to the London and North Eastern Railway Co., which came into existence on January 1, 1923, by the amalgamation of the North Eastern, Great Central, Great Eastern, Great Northern, North British, and Great North of Scotland Railway Companies. In tracing the history of the railways in the riding we may begin with the Hull and Selby Railway Co., which obtained an Act of Parliament in 1836. Its line, 30¾ miles in length, was opened in 1840, and, being connected with the Leeds and Selby Railway, opened in 1834, forms one of the most important trade routes in the country. It traverses a very level stretch of country, the only noteworthy cutting being that at Hessle Cliff. It crosses the Ouse at Selby, by an iron bascule bridge, with a movable arch of two flaps, allowing a clear waterway of 45 feet. The tickets issued in the early days of this railway were of paper, on which and their counterfoils the destination and fare were written by hand. The station in Hull was on the west side of the Humber Dock and was approached from Railway Street.

The York and North Midland Railway Co. was formed in 1835 and in 1845 it completed its line from York to Scarborough, 42¼ miles, with a branch of 6½ miles from Rillington to Pickering. After crossing the Ouse and leaving York, this line does not enter the East Riding until Huttons Ambo station, where it originally crossed the Derwent by a timber bridge 400 feet long. After 18½ miles it leaves the riding 2¼ miles short of Seamer station. The York and North Midland were authorised, in 1845, to construct a line from Seamer to Bridlington, of which the branch from Seamer to Filey, 6½ miles, was opened in 1846 and the continuation to Bridlington, 13½ miles, in 1847. The same company opened in 1846 the line from Hull to Bridlington, 31 miles, *via* Beverley and Driffield; in 1847 the line from York to Market Weighton, 21½ miles; and in 1848 the line from Selby to Market Weighton, 16¼ miles.

In 1848 a new (Paragon) Station was built at Hull, in Paragon Street, and connected with the Hull and Selby and Hull and Bridlington lines and, in 1864, with the lines to Withernsea and Hornsea.

The Malton and Driffield Junction Railway Co.

obtained an Act in 1846 and opened their line of 19 miles in 1853. It includes the Burdale Tunnel under the Wolds, 1734 yards long. In 1853 the Hull and Holderness Railway Co. obtained an Act for a railway from Hull to Withernsea, 17½ miles, which was opened in 1854. Their Hull station was where the Drypool Goods station is now. "Under the powers of an Act of Parliament of 1854 the York and North Midland and Leeds Northern Companies were dissolved and their undertakings vested in the York, Newcastle, and Berwick Co., who now assumed the comprehensive title of the North Eastern Railway Co."

The Hull and Hornsea Railway Co. obtained their Act in 1862 and opened their line, 13 miles, in 1864. The line from Market Weighton to Beverley was opened in 1865. Although an Act had authorised a line from Thorne through Goole to a junction with the Hull and Selby line at Staddlethorpe in 1863 it was not opened until 1869*. The Scarborough, Bridlington, and West Riding Junction Railway Co. obtained an Act in 1885 for a line from Market Weighton to Driffield, which was opened in 1890 and completed the connection between Filey, Bridlington, and the West Riding *via* Selby.

The Hull and Barnsley Railway Co. obtained an Act in 1880 and its main line from Hull (Cannon Street and Beverley Road) to a junction with the Midland Railway at Cudworth, near Barnsley, was opened in 1885. Between Little Weighton and South Cave it passes under the Wolds in Drewton Tunnel, 2116 yards, and runs *via*

---

* The main line from Selby to York was opened as a whole in 1871 and the Derwent Valley Light Railway from York to Cliff Common in 1913.

Howden to Barmby, between which and Drax it crosses the Ouse by a lattice girder bridge of three spans, of which the central one is a swinging span of 100 feet.

All the lines previously mentioned were absorbed, at various dates, by the North Eastern Railway Co., which has now become, by amalgamations, as previously mentioned, the London and North Eastern Railway Co. The number of persons employed on railways in Hull in 1911 was 5714 men and 28 women and in the Administrative County of the East Riding 1763 men and 25 women. In York 3095 men and 73 women were employed, besides 1258 Railway-Coach and Waggon Makers.

25. Administration and Divisions— Ancient and Modern.

The division of Yorkshire into ridings, which are so arranged that all three meet at York, probably occurred in the latter half of the ninth century, when Danish kings reigned at York. The meeting-place of the *thing* or court of the East Riding was at Craikhow, near Windyates, in the no longer inhabited Gardham on the Wolds, between Beverley and Market Weighton. For centuries the riding has been divided into wapentakes. The word " wapentake " derived from the Norse, means literally "weapon taking " or " weapon touching " and was applied to the form of ratifying the decisions of the local court. It is curious that although the wapentake is considered specially characteristic of the Danes and the East Riding is the most Danish part of Yorkshire, yet in the Domesday Survey (1086) the East Riding divisions (except Holderness), are not wapentakes but "hundreds," smaller than wapentakes. Dr W. Farrer states that the hundreds were grouped into wapentakes in the time of Henry I. Although the division into the wapentakes of Buckrose, Dickering, Harthill, Holderness, Howdenshire, and Ouse and Derwent persisted into the nineteenth century these divisions are no longer used for any purpose.

The Shire Moot (which was not only a court but an assembly of the people) was held twice a year—that is the full court—but according to the charter of 1217 it met once a month. Its constituting official was the shire-reeve or sheriff, who was accompanied by the ealdorman (called earl under the Danish kings) or governor of the shire, and the bishop, who respectively declared the secular and the spiritual law. In 1340 it was provided by statute that no sheriff should continue in office for more than a year. In the eighteenth and nineteenth centuries the practical power of the sheriff was further diminished. Even by 1689 he discharged his functions by deputy, appointing a professional Under Sheriff, usually the Clerk of the Peace and the same person who had served his predecessor in office. The chief duty of the sheriff at present is to receive the Judges of Assize with ceremony. There is only one High Sheriff for Yorkshire. The office of Lord Lieutenant is a life appointment and dates only from the middle of the sixteenth century; from the reign of William and Mary it was usually combined with that of *Custos Rotulorum* or "Keeper of the Rolls of the Peace." The Lord Lieutenant appointed the officers of the militia and took command in an emergency of all the local forces of the county. The Justices of the Peace are appointed upon his

nomination and he is president of the Territorial Association of his county. During the eighteenth and nineteenth
centuries the local government of counties fell largely
into the hands of the Justices of the Peace in Quarter
Sessions. By the Local Government Act of 1888 County
Councils were established and took over the functions of
local government previously fulfilled by the justices, with
additional functions. Each riding of Yorkshire is a separate administrative county, with a separate county council.
The East Riding County Council consists of 52 councillors, representing 52 electoral divisions with one member
each although there are four divisions in Beverley and
three in Bridlington. They hold office for three years.
There are, in addition, 18 aldermen, who are elected by
the councillors and hold office for six years. Beverley is
the shire town and meeting-place of the East Riding
County Council. The administrative county is divided
into districts which are either urban or rural. Urban
districts include boroughs. By the Act of 1888 boroughs
were divided into three classes : (1) The County Borough,
i.e. boroughs already counties of themselves or having a
population of not less than 50,000. They are completely
independent of the County Council but have all its powers.
Kingston-upon-Hull is a county borough. Its government
is vested in a Corporation consisting of a Lord Mayor,
16 Aldermen, 48 Common Councillors, a Recorder and
a Sheriff; (2) The larger Quarter Sessions Boroughs, with
a population exceeding 10,000. These are subject to the
control of the County Council for certain special purposes
only. (3) The boroughs with a population of under
10,000. Subordinate to the County Council there are

in the East Riding three municipal boroughs (places which have been incorporated by royal charter), viz. Beverley, Bridlington, and Hedon; eight other urban districts, viz. Cottingham, Filey, Great Driffield, Hessle, Hornsea, Norton, Pocklington, and Withernsea; and twelve rural districts, viz. Beverley, Bridlington, Driffield, Escrick, Howden, Norton, Patrington, Pocklington, Riccall, Sculcoates, Sherburn, and Skirlaugh. By the Local Government Act of 1894 Parish Councils were established, of which there are 131 in the East Riding. There are 338 civil parishes or townships in the riding. The parliamentary representation of the East Riding is as follows: Parliamentary Borough, Kingston-upon-Hull, returning four members; and three county divisions, named respectively Buckrose, Holderness, and Howdenshire, each returning one member.

Ecclesiastically the East Riding is in the diocese of York. Hull was from 1534 to 1553, and is since 1891, the seat of a suffragan bishop. Beverley is also the seat of a suffragan bishop.

26. The Roll of Honour.

The East Riding and the city of York are the birth-place of many distinguished men. Among eminent theologians and divines St John of Beverley, believed to have been born at Harpham, was consecrated bishop of Hexham in 687 and was bishop of York from 705 to 718. He founded a monastery at Beverley, where he died in 721. He was canonised in 1037 and many miracles were attributed to him. Alcuin (Albinus), the most distinguished scholar of the eighth century, the confidant and adviser of Charlemagne, was born at York in 735. He wrote, *inter alia*, a history in verse of the church at York. He became abbot of St Martin at Tours, where he died in 804. John of Bridlington was born at Thwing in 1320 or 1324, became prior of Bridlington in 1361, died 1379, and was worshipped as a saint within a few years of his death. Walter Skirlaw, bishop successively of Lichfield, Bath, and Durham, was born at South Skirlaugh and was distinguished by his liberality in building Skirlaugh Chapel, the chapter-house and central stage of the tower at Howden, and other architectural works. He died in 1406. John Fisher, born at Beverley about 1459, became bishop of Rochester and Chancellor of the University of Cambridge. For refusing to acknowledge Henry VIII as supreme head of the church, he was beheaded in 1535. Thomas Lamplugh, born at Octon near Thwing in 1614 (d. 1691), was appointed bishop of Exeter and in 1688 archbishop of York. He was a liberal benefactor of York Cathedral. Richard Osbaldeston, born at Hunmanby in 1690 (d. 1764) was successively bishop of Carlisle and of London. He prohibited the introduction of statuary at St Paul's. John Green was born at or near Hull about 1706 (d. 1779) became bishop of Lincoln and was vice-chancellor of Cambridge University in 1756.

In the fourteenth and fifteenth centuries it was not uncommon for divines to be also statesman. For instance John Hotham, born at Scorborough (d. 1337) was bishop of Ely and twice Lord Chancellor, and John Alcock, born 1430 at Beverley (d. 1500) was successively bishop of

Rochester, Worcester, and Ely and twice Lord Chancellor. He made the Hull Grammar School a Free School.

Sir Michael de la Pole, eldest son of that Sir William de la Pole whose tomb is in Holy Trinity Church, Hull, was born about 1330 and was a favourite of Richard II, who made him Lord Chancellor, admiral, and Earl of Suffolk. He was found guilty of peculation and died in exile in Paris in 1389. Mention has already been made in Chapter 17 of the rebel Robert Aske, who was probably born at Aughton and was hanged in 1537. Sir William Strickland of Boynton, born about 1596 (d. 1673) was a stout puritan who represented Hedon in the Long Parliament and was summoned by Cromwell to his House of Lords. Sir William's younger brother, Walter Strickland (flourished 1640–1660) was in 1642 appointed by the Long Parliament their agent in Holland and was a member in both the Councils of State during the Protectorate. Sir James Hudson, born at Bessingby Hall, Bridlington, 1810 (d. 1885) was British envoy at Rio de Janeiro in 1850, minister at Turin 1851 to 1863, and was awarded the Grand Cross of the Bath in the latter year. Thomas Ward, born at Howden in 1810, became minister to the Duke of Lucca (afterwards Duke of Parma) and died in 1858. Sir John Hall was born at Hull in 1824 (d. 1907). He emigrated to New Zealand in 1852, was elected to the Colonial Parliament in 1855, became Premier in 1879 and was awarded the K.C.M.G. in 1882.

Although primarily a journalist rather than a politician Sir W. C. Leng, born at Hull in 1825 (d. 1902) may be mentioned here. He became editor of the *Sheffield Daily*

Telegraph and in 1867 denounced trade-unionist terrorism in Sheffield at personal risk.

Among soldiers, Sir Marmaduke Constable of Flamborough, born about 1455 (d. 1518), took part in wars in France in 1475 and 1492, and in the battle of Flodden in 1513. Although hardly to be included in a Roll of Honour, Sir John Hotham, who was created a baronet in 1621 and was repeatedly M.P. for Beverley and in 1641 Governor of Hull, has been already mentioned on page 90. Sir William Constable of Flamborough raised a regiment of foot for the Parliament. In 1644 he took Bridlington, assisted in the capture of Whitby, re-took Scarborough, defeated Newcastle's forces at Driffield and Malton, and captured Tadcaster and Stamford Bridge. He died in 1655.

On the Royalist side Sir Philip Monckton, born about 1620 at Cavil near Hedon (d. 1679) was a captain at the first siege of Hull and distinguished himself at Atherton (Adwalton) Moor, Marston Moor, and Naseby.

The East Riding has produced several ancient chroniclers, among whom was William of Newburgh, born in 1135 at Bridlington (d. 1200), who wrote *Historia rerum Anglicarum*, covering the period 1066–1198. Roger of Hoveden flourished 1174 to 1201 and was probably a native of Howden. His *Chronica* extends from A.D. 732 to 1201, of which the portion from 1192 to 1201 is a primary authority. Walter de Hemingburgh, who was sub-prior of St Mary's, Gisburne (Guisborough), flourished 1300. His *Chronicle* extending from 1066–1297 continued to 1346 (but 1316–26 missing) and is "one of the most favourable specimens of our early chronicles." Peter of

Langtoft, who died 1307, may have been born at Langtoft. He wrote a *Chronicle* in French verse from the earliest period to the death of Edward I.

The East Riding is the birthplace of several noteworthy antiquaries and topographers. The Rev. W. Dade (1740 ?–1790) was born at Burton Agnes. In 1783 he published *Proposals for the History and Antiquities of Holderness* but never completed the undertaking. It was taken up and completed by George Poulson (1783–1858) who in 1829 published *Beverlac ; or the Antiquities and History of the Town of Beverley*, and in 1840–1 *The History and Antiquities of the Seigniory of Holderness*. John Bigland (1750–1832), born at Skirlaugh, wrote articles and 16 books, among which was *A Topographical and Historical Description of the County of York*, 1812. Thomas Thompson, M.P., F.A.S. (1753–1828), born at Owbrough, wrote *Ocellum Promontorium or Short Observations on the Ancient State of Holderness*, 1824, and a *History of the Church and Priory of Swine*, 1824. Charles Frost 1781 (or 1782)–1862 was born at Hull. In 1827 he published *Notices Relative to the Early History of the Town and Port of Hull*. The Rev. Marmaduke Prickett, born at Bridlington in 1804, was the author of *An Historical and Architectural Description of the Priory Church of Bridlington*, 1831. No one did as much to elucidate the prehistoric antiquities of the riding as the late J. R. Mortimer (1825–1911), born at Fimber. He published 47 papers on archæological and geological subjects, and in 1905 his chief work, *Forty Years Researches in British and Saxon Burial Mounds in East Yorkshire*, with 125 plates. Mr J. R. Boyle, born at Accrington, Lancashire, in 1853 (died in Hull 1907)

Andrew Marvell, from a miniature
(*Reproduced by permission of His Grace the Duke of Buccleuch*)

published in 1889 *The Lost Towns of the Humber*, and more recently *The Early History of the Town and Port of Hedon*.

Compared with its wealth in chroniclers and topographers the East Riding has produced few poets or writers of fiction. One of the most distinguished natives of the riding was Andrew Marvell (1621–1678), born at Winestead, who became assistant secretary to Milton and member of Parliament for Hull from 1659 until his death. He wrote exquisite poetry, especially on gardens, and political satires in prose and verse, and was remarkable in that age for his incorruptibility. William Mason (1725–1797), born at Hull, became rector of Aston near Rotherham. He was a friend of the poet Gray and himself wrote odes and plays. Among novelists, Mrs Stannard (John Strange Winter), 1856–1911, born at York, is known as the author of some sixty light and amusing books. York is the birthplace of John Flaxman (1755–1826), the eminent sculptor and draughtsman, who was employed by Josiah Wedgwood as a modeller of classic and domestic friezes and executed many monuments. York also claims William Etty (1787–1849), a great master of flesh tints, who has few equals as a colourist. Hull is the birthplace of John Bacchus Dykes (1823–1876), who received the honorary degree of Doctor of Music from Durham in 1849 and composed many of the hymn tunes in *Hymns Ancient and Modern*, while York can claim Sir Joseph Barnby (1838–1896), the musical composer and conductor. He wrote an oratorio and 246 hymn tunes.

Two distinguished entomologists were born at Hull, viz. A. H. Haworth (1767–1833) author of *Lepidoptera Britannica* and a botanical work *Synopsis Plantarum Succu-*

lentarum, and William Spence (1783–1863), joint author with W. Kirby of their *Introduction to Entomology*, 1815. Edward Tindall (d. 1877) of Bridlington, wrote on the Chalk and on the Bridlington Crag. Hugh E. Strickland (1811–1853), ornithologist and geologist, born at Reighton, wrote on the geology of Asia Minor and on the Dodo.

Agriculture owes a great debt to the chemist, Sir J. H. Gilbert (1817–1901), born at Hull, who was associated with J. B. Lawes in the Agricultural Experimental Station at Rothamsted. Perhaps the most distinguished of Yorkshire mycologists was George Massee, F.L.S. (1850–1917), born at Scampston, author of 250 books and papers, chiefly on Fungi.

Among medical men Sir B. F. Outram, F.R.S., K.C.B., F.R.C.P. (1774–1856), born at Kilham, became in 1841 medical inspector of Her Majesty's fleets and hospitals; Sir James Alderson, M.D., F.R.S., born at Hull 1794 (d. 1882), was elected President of the College of Physicians in 1867, knighted in 1869, and appointed Physician Extraordinary to the Queen in 1874; Humphrey Sandwith, C.B., D.C.L. (Oxon.) (1822–1881), born at Bridlington, accompanied Layard to Nineveh, was Inspector-General of Hospitals in the Russo-Turkish War of 1855, and distinguished himself during the siege of Kars.

Sir Christopher Sykes, Bart. (1749–1801), born at Roos, and his son Sir Tatton Sykes (1772–1863) have been mentioned in Chapter 12. The latter was well known for his horse- and sheep-breeding, and as a patron of the turf; his son as a benefactor of the National Church.

In shipping circles no name is better known than that of Charles Henry Wilson (1833–1907), first Baron Nunburnholme (1905), born at Hull. He became joint manager with his brother, Arthur Wilson (1836–1909), of Thomas Wilson, Sons & Co., the largest private ship-owning firm in the world.

Of all her citizens perhaps Hull is proudest of William Wilberforce (1759–1833), philanthropist. He was elected M.P. for Hull in 1780, became a close friend of Pitt, who

William Wilberforce

in 1788 introduced the subject of the slave trade in Parliament. Wilberforce continued to agitate for the abolition of slavery and the Emancipation Bill was passed a month after his death. He is commemorated by a column in Victoria Square, Hull, and his Hull residence has been bought by the Corporation and converted into a Museum.

27. The Chief Towns and Villages of the East Riding.

The figures in brackets give the population in 1921. The other figures are references to the pages in preceding chapters. Abbreviations: Sax.=Saxon, Norm.=Norman, Trans.=Transitional Norman, E.E.=Early English, Dec. =Decorated, Perp.=Perpendicular.

Owing to the census of 1921 being taken in June the excess of census over estimated *resident* population expressed as a percentage of the former was in Bridlington M.B. 32·2, Filey U.D. 25·9, Hornsea U.D. 13·5, Withernsea U.D. 17·7.

Aldbrough (743). The nave arcades and W. tower of church Trans., windows of aisles and clerestory Perp. Altar tomb of Sir John de Melsa (Meaux), d. 1377. Saxon sundial, p. 103. Former castle, p. 125. Brewery, p 71. Impassable road, p. 139. Coast erosion, p. 48, Fishing boats, p. 77.

Anlaby (1986). Eccl. parish, 1 mile S.E. of Kirk Ella, formed in 1902 from the parishes of Cottingham Hessle and Kirk Ella. The Gothic brick church built in 1864 was rebuilt and enlarged in 1885. Date of road from Hull, p. 138.

Bainton (309). In the chancel of the beautiful Curvilinear (Dec.) church is the brass of Roger Gudale, Rector 1382–1429, and in the S. aisle the wall tomb of Sir Edmund de Mauley, who fell at Bannockburn in 1314.

Beverley (13,469) is a municipal borough, shire town of the East Riding, seat of a suffragan bishop. "Around the great and famed shrine of St John of Beverley the town of Beverley sprang up, a minster town exclusively in its earlier centuries." Leland writing 1535–1543 describes the town as "buildid of wood." The roads to Beverley, with the exception of that from Hull, were all formed as roads to the minster. Sanctuary rights were formally accorded by Athelstan in 937 and Beverley Minster became the greatest of England's historic sanctuaries. The sacred stone chair or Frith Stool has been mentioned, p. 125. In the British Museum is a register of those fugitives to Beverley, between the years 1478 and 1539, who sought perpetual immunity. They numbered 439, of whom 208 were debtors and 186 were guilty of homicide or manslaughter. The Act 32 Henry VIII (1540) extin-

guished all special rights of sanctuary. Beverley first returned
two burgesses to Parliament in 1295. It was disfranchised in 1870.
The Grammar School has 180 boys. The most attractive features
of Beverley are the Minster, p. 123, and St Mary's Church, p. 111.
It is disputed whether the town was walled, but it had five gates or
bars, of which North Bar remains, p. 125. There are two squares
called respectively the Wednesday and Saturday Markets. In the
latter is the octagonal Market Cross, built 1714 and repaired
1769. Remains of Dominican Friary, p. 118. Geology, p. 26.

**Saturday Market Place with Market Cross and St Mary's
church tower, Beverley**

Chariot burial, p. 96. History, pp. 88, 99. Industries, pp. 68-71.
Shipping and trade, p. 79. Roads, p. 138. Canal, p. 141. Railways, pp. 143, 144. County Council, p. 147. Natives, p. 149.

Bishop Burton (378) is one of the most picturesque villages in the East Riding. The W. tower of the church is E.E.
and has doorways on N., S. and E. Under mats in the chancel are
the chalice brass of Peter Johnson, vicar (d. 1460) and other
brasses. Roman tesselated pavements, p. 101. Jacobean Manor
House, p. 135.

Bishop Wilton with Belthorpe (465). The village lies at the foot of the loftiest part of the Wolds. The houses are ranged on each side of a sparkling beck with sloping grassy banks. A rectangular dry moat surrounds the site of a former Palace of the Archbishops of York. The beautiful church, see p. 108.

Brandesburton (577). The nave arcades and chancel of the church are E.E. In the chancel is the bracket brass of William Darell, rector of Halsham and Brandesburton (d. 1364) with his half-length headless figure and the large (headless) brass of Sir John St Quintin (d. 1397) and Lora, his wife. In the village is an ancient cross.

Church and village of Bishop Burton

Bridlington (22,768) has been a municipal borough since 1899. It consists of two portions, which are however continuous with one another—Bridlington Quay, the popular modern watering-place and the Old Town, a mile N.W. of it. Old Bridlington is "a rather sleepy place, with some of the characteristics of a market town and some resemblance to a cathedral city." The only features of interest are the Bayle Gate, p. 126, and the beautiful Priory Church, p. 117, which consists of the nave and W. towers only of the original church, the rest having been destroyed in 1539.

Bridlington Quay has north of the harbour, p. 46, four spacious promenades at varying levels. The chief one, called the Royal

Prince's Parade, has a Grand Pavilion for concerts. South of the harbour are the New Spa Gardens with the Spa Theatre and Opera House. The Grammar School has 386 boys. The sands are excellent; boating, bathing and fishing are much indulged in and Flamborough Head is an attractive feature. Geology, pp. 25, 30. Gipsey Race, p. 16. Coast, p. 46. Fisheries, pp. 77, 78. Industries, pp. 68, 69. Communications, p. 143. History, p. 90. Natives, pp. 151, 152, 155.

Bubwith (469) is situated on the left bank of the Derwent, here about 60 feet wide. The handsome church has an E.E. nave and chancel and a Norman chancel arch, recessed in three orders. The Perp. W. tower was erected about 1424 by Nicholas de Bubwith, Bishop of Bath and Wells, a native of the village.

Cave, North, with Everthorpe and Drewton (1019).
The nave, clerestory and upper part of W. tower of the cruci form church are Perp., the nave arcades Trans. St Austin's Stone on the Wolds, 2½ miles N.E. by E. of the church consists of a breccia of chalk flints cemented together and is purely natural.

Cave, South (974).
The church, except the Perp. tower, was rebuilt in 1601; the chancel was rebuilt in 1846–7 and the nave was restored in 1848. Roman pig of lead found here, p. 97.

Cottingham (5133), urban district and civil parish. Here is the site of Baynard's Castle, p. 126. A mile and a quarter S. of Cottingham is the site of the Augustinian Priory of Haltemprice, founded by Thomas Wake in 1322. The Hull Corporation constructed waterworks at Springhead in this parish in 1862 and at Milldam in 1889, these places being about three miles apart. The water is obtained from wells in the Chalk. Church, p. 114.

Driffield, Great (5674), urban district and C.P., is still essentially an agricultural town, its corn market being one of the largest in the East Riding. Industries, pp. 68, 70. Canal, p. 141. Railways, pp. 143, 144. Prehistoric remains, p. 95. Anglo-Saxon remains, p. 102. Former castle, p. 126. Church, p. 110.

Easington (336). The nave arcades of the church are E.E., but the W. tower is Perp. Beautiful E.E. blocked doorway on N. side of N. aisle. Near the church is a thatched tithe barn believed to be of the fourteenth century. Coast erosion, p. 49.

Elloughton with Brough (1230). Elloughton church has a fine E.E. south doorway. Brough is the site of a Roman ferry across the Humber, p. 96.

Emswell (or Elmswell) with Little Driffield (301). At both these places springs rise in the Chalk, which form trout streams and are among the sources of the River Hull. The chancel of Little Driffield church is reputed to be the burial place of Alfred, King of Northumbria (d. 705). The E. arch of the W. tower consists of two arches, one within the other, and is early Norman. Pre-Conquest sculpture, p. 103.

Escrick (597). The church was burnt down in 1923. Escrick Park, the seat of Lord Wenlock, contains about 459 acres of rich pasture land, well wooded and stocked with 80 fallow deer.

Filey (4549) urban district and C.P. is a quieter, more select and exclusive watering-place than Bridlington. It lies just over 100 feet above sea level, on the top and seaward slopes of a cliff of Boulder Clay in a beautiful bay, bounded by Filey Brig, p. 41, on the north and Flamborough Head, p. 43, on the south. It is protected by a strong sea wall, on the top of which is a promenade, p. 43. The cliff slopes and the Church Ravine, originally eroded by Filey Beck, are beautifully wooded. This ravine formerly separated the East from the North Riding, p. 5; the parish church of St Oswald, p. 110, then being in the North Riding. The best part of the town is the Crescent on the sea-front. In the Crescent Gardens are the Roman pedestals mentioned on p. 100. The sands are among the best in England, p. 43. Bathing, boating and fishing are favourite amusements. Geology, pp. 10, 20, 23. Coast erosion, p. 48. Fisheries, pp. 77, 78. Communications, pp. 143, 144.

Fimber (173), 2½ miles W.N.W. of Wetwang. The church was rebuilt in 1871 by Sir Tatton Sykes. Birthplace of J. R. Mortimer, the archæologist, p. 152. He has described ancient hollow ways (sunk roads or covered ways) older than the Wold entrenchments, p. 95, and has traced six of them radiating from Fimber.

Flamborough (1325) has an interesting church, p. 108, and the scanty remains of the misnamed Danish tower, p. 126; but the geology, p. 22, coast scenery and lighthouse, pp. 43–46, birds, p. 38, and misnamed Danes' Dyke of Flamborough Head, p. 95, are the chief attractions. Coast erosion, p. 48. Fisheries, pp. 77, 78. History, p. 91. Natives, p. 151.

Fridaythorpe (242). The W. tower, chancel arch and S. doorway of nave are Norm. or Trans. Font Norm. North aisle E.E. Roman road, p. 98.

Frodingham, North (543). The church has a W. tower partly Perp. The nave and chancel, rebuilt in 1874, appear to have been E.E. At the vicarage is a pre-Conquest cross head, p. 103.

Goodmanham (268). For the place name see p. 62. Church, p. 106. Bronze age cinerary urn, p. 94. History, p. 86.

Grimston, North (124). The aisleless nave and W. tower of the church are E.E.; the chancel, chancel arch and the great cylindrical font sculptured with the Last Supper, Descent from the Cross and a figure of (?) St Nicholas are Norm. Geology, p. 19. Minerals, p. 74. Prehistoric antiquities, p. 96.   •

Harpham (221), 1¼ miles S.W. of Burton Agnes. An almost unique case of obtaining a royal license to fortify the belfry of a church is that of Harpham in 1374. The church contains the altar tomb of William St Quintin (d. 1349) and his wife; the double brass of Thomas St Quintin and his wife 1418 and a smaller brass of another Thomas St Quintin (d. 1445). Roman remains, p. 100. Reputed birthplace of St John of Beverley, p. 148.

Hedon (1321), a municipal borough, with the oldest civic mace in England, dating from the reign of Henry VI, was formerly a rival of Hull as a seaport, p. 79, but is now only noteworthy for its beautiful church, p. 111. Hedon first sent members to Parliament in 1295, next in 1547 until disfranchised in 1832. Edward III in 1348 granted the burgesses the privilege of electing a mayor and bailiffs every year, and the town was incorporated by Charles II in 1661. The beautiful cross which formerly stood at Kilnsea (supposed to commemorate the landing of Henry IV or Edward IV) has been removed to Hedon, where it stands in front of Holyrood House. Hedon has a manufacture of bricks and a trade in corn. The turnpike road connecting it with Hull was constructed in 1832, see p. 138.

Hemingbrough (529). The church was made collegiate in 1426 by the Prior and convent of Durham, and it remained so until the dissolution in the reign of Edward VI. See p. 125. A great meander of the Ouse has been artificially cut off here.

Hessle (6107) urban district and C.P. lies at the southern end of the Wolds and has become a suburb of Hull. The church has

The Old Kilnsea Cross as re-erected at Hedon

a nave, of which the 3 western bays of the arcade are E.E. as is the beautiful rebuilt south doorway, recessed in 4 orders; the W. tower, within the aisles, is Perp. Chancel rebuilt 1868–70.

Geology, p. 26. There are large chalk quarries and two whiting factories. Shipbuilding, p. 69. Road to Beverley, p. 138 and from Hull, p. 138.

Holme upon Spalding Moor (1596). The church is splendidly situated on the top of a hill 150 feet high ½ mile E. of the village. Geology, p. 9.

Hornsea (4278) urban district and C.P. This seaside watering place is described, p. 46. The parish church's W. tower (of uncertain age) is within the aisles, the outer walls of which are E.E. Late in the fifteenth century the centre of the church was pulled down and the present Perp. nave, chancel and S. porch were built. In the S.W. corner of the churchyard is the old market cross, and another ancient cross stands in Southgate. Hornsea Mere is described p. 17. William, the eleventh abbot of Meaux, in 1260 claimed the right to fish in the south part of Hornsea Mere, a claim opposed by the abbot of St Mary's, York. The controversy was decided against Meaux by combat between representative champions. Geology, p. 30. Coast erosion, pp. 48, 53. Anglo-Saxon antiquities, p. 102. Birds, p. 37. Fisheries, pp. 77, 78. Railway, p. 144.

Howden (2052). The church of St Peter at Howden, see p. 125, was given at the Conquest to the Prior and Convent of Durham. The Bishops of Durham were lords of the Liberty of Howden which, according to the Nomina Villarum of 1315–16, comprised more than 50 villages and was not co-extensive with the wapentake of Howdenshire. Bishop's Manor House, p. 132. Natives, pp. 150, 151.

Hull, *officially* **Kingston-upon-Hull** (287,013; in 1801 only 29,849), a city and county of a city, county and parliamentary borough, returning four members to Parliament, seat from 1534 to 1553 and since 1891 of a suffragan bishop, is the third port of the United Kingdom.

Hollar's view of Hull in 1640, see p. 129, shows that at that time the city was confined to the area now bounded by the Queen's, Prince's, and Humber Docks and the rivers Humber and Hull, except that the castle and blockhouses were on the east bank of the Hull. At the present time Victoria Square has on its W. side the Art Gallery and City Hall in Renaissance style with a lofty dome; on its E. side the handsome Dock Offices with three domes; on its S. side the column forming the Wilberforce Monu-

Humber, Prince's, and Queen's Docks and Victoria Square, Hull

ment. The new Guildhall and Law Courts in Alfred Gelder Street form a fine block also in Renaissance style. Hull Trinity House, built in 1753, in Tuscan style, is the home of an institution founded in 1369. Hull is well provided with municipal Museums. The chief one is the Royal Institution building, Albion Street, and comprises a good natural history collection and an excellent collection of antiquities, including the Mortimer collection of prehistoric remains. The Wilberforce House is a museum of historic objects and the Museum of Fisheries and Shipping in Pickering Park is of interest. The city is also well provided with parks, viz.: Pearson Park, 25 acres; West Park, 31 acres; East Park, 70 acres; Pickering Park, 50 acres, besides recreation grounds.

Among numerous schools Hymers College, a first grade school for boys, attendance over 450; the new Hull Grammar School (built 1891) with about 500 boys; the Newland High School for girls, attendance about 500; the Technical College, Park Street, and the School of Art may be mentioned. Hull has its own Municipal Telephone System. The city occasionally suffers from floods, due to high tides. On December 17, 1921, a flood did damage estimated at £100,000. Geology, see p. 31. Climate, p. 56. Industries, pp. 68–71. Fisheries, pp. 75–78. Shipping and Trade, pp. 80–84. History, pp. 88–91. Churches, pp. 113, 114. Former fortifications, p. 128. Domestic architecture, pp. 132, 134, 137. Roads, p. 138. Railways, pp. 142–145. Corporation, p. 147. Natives, pp. 149, 150, 152, 154, 155, 156.

Hunmanby (1501). The chancel arch and base of W. tower of the church are Norm., the nave arcade E.E. In the market-place is the shaft of an ancient cross. There are two brick works. Chariot burial, p. 96. Site of castle, p. 127. Pre-Conquest cross head, p. 103. Native, p. 149.

Hutton Cranswick (961) consists of two villages with the church at Hutton. Its nave arcade and chancel arch are E.E., south doorway Norm., W. tower Perp. The original Norman font is in York Museum (Hospitium), see p. 110.

Keyingham (595). In the church the piers of the Dec. (?) nave arcade consist of 4 clustered shafts only 4 feet 6 inches high between base and necking. The spire of the W. tower resembles that of timber churches. Arms landed in Keyingham Creek in 1642, p. 90.

Kilham (853). In the middle of the eighteenth century Kilham was the chief market town of the Wolds, but it has been superseded by Great Driffield. Church, p. 108. Native, p. 155.

Kilnsea (139). Many Roman antiquities have been found in this parish from time to time. Coast features, p. 47. Coast erosion, pp. 49, 52. The ancient cross which stood at Kilnsea has been removed to Hedon.

Kirkburn and Battleburn (92). Kirkburn has a very fine Norm. church, p. 106, and an interesting Norm. font sculptured with figures.

Kirk Ella (491) consists chiefly of country residences embowered amongst trees. The church has an E.E. nave and chancel and a handsome Perp. W. tower.

Kirkham (26). Here are the ruins of an Augustinian Priory on the banks of the Derwent, amongst the most beautiful scenery of the riding, p. 117. Ironstone, p. 74.

Langtoft (470). Church, p. 111. The village is subject to floods; a great one occurred in 1657 and another on July 3, 1892, when the village street was 7½ feet under water and a large gully was torn in the side of the dale. Native, p. 152.

Lockington (448). The church has a Norm. S. doorway and remains of a Norm. chancel arch, a Dec. chancel and on S. of nave a chapel panelled with 173 coats of arms of the Constable family. At ¼ mile S. of the church is a motte, p. 127.

Londesborough with Easthorpe (297). The church has pre-Conquest sculpture in the tympanum of Norman S. doorway, p. 102. Nave arcade, font and lower part of tower E.E. Monuments of the Cliffords and a vault of the Earls of Burlington. Roman road, p. 97.

Market Weighton and Arras (1717) is a market town situated opposite a gap at the western foot of the Wolds. The Roman road called Humber Street runs ½ mile E. and another Roman road ran 1 mile W. of the parish church of which the base of the tower is probably early Norm. and the N. nave arcade E.E. or Trans. The font is Norm. Geology, pp. 20, 21. Minerals, p. 73. Industries, p. 68. Canal, p. 141. Railways, pp. 143, 144.

Middleton on the Wolds (628). The church has a Dec. nave, E.E. chancel and Trans. font.

Naburn (556), on the Ouse, has a church built in 1854 and a Maypole. Naburn Lock, p. 139, is ¾ mile S. of the village.

Nafferton (1237). The church is chiefly Dec. or Perp. and has a Norm. font.

Newbald, North (566), lies in a gap at the W. foot of the Wolds and possesses the finest Norm. church in the riding, p. 106.

Norton (3854). The new parish church was partially erected and opened in 1894; additional portions in 1911; completed in 1913. Norton and its neighbourhood are noted for training and breeding establishments for racehorses. In 1862 a section of a Roman road was exposed at Norton.

Ottringham (516). The church has a W. tower, upper part Dec. with a parapetted broach spire. E. arch of tower Trans. pointed, with chevron and rolls. Coast erosion, p. 50.

Patrington (1137) has a beautiful Dec. church, p. 111. A navigable branch of the River Humber formerly adjoined part of the town called Havenside, but it was closed in 1869 and is now used for drainage only.

Paull (540) is on the left bank of the Humber. The cruciform church is Perp. Near the river is a battery, rebuilt in 1857 and remodelled in 1894. Paul Holme Tower, p. 134.

Pocklington (2645). Church, see p. 111. Pocklington Grammar School was founded or at least endowed by John Dowman in 1514 and in 1921 there were 134 boys in the school, mostly boarders. The local industries include brewing, malting, rope and twine making, agricultural implement works and corn mills. Pocklington Canal, p. 142.

Preston (1422). The handsome W. tower and the clerestories of the church are Perp.; the S. arcade of nave E.E. In S. aisle interesting sculpture in alabaster discovered in 1880 buried near the pulpit is either an Easter Sepulchre or an Altar Piece.

Riccall (739). West tower and S. doorway of church Trans., nave arcades E.E. except eastern bay which is Dec. Landing of Tostig and Harald Hardrada in 1066, see p. 87.

Rudston (492). Church, p. 107. Prehistoric monolith, p. 93. Long Barrow, p. 92. Roman road, p. 99. Roman pavement, p. 100.

Settrington (448), 3½ miles E.S.E. of Malton. The village lies at the foot of the Wolds, the church and hall embowered in fine trees. The S. doorway and possibly the nave arcade of the church are Trans., W. tower Perp.

Sherburn (581). The church has a Norm. chancel arch and a very fine Norm. font. The E. arch of tower is E.E. and the chancel Dec. Pre-Conquest sculpture, p. 103.

Skipsea (286) was the capital seat of the barony of Holderness in the reign of the Conqueror. Site of castle, p. 127. Coast erosion, pp. 48, 50.

Sledmere (491), the seat of the Sykes family, see pp. 65, 155. The beautiful church, with the exception of the tower, was erected by the second Sir Tatton Sykes. On Garton Hill 2½ miles S.E. is a tower monument to the first Sir Tatton Sykes.

Stamford Bridge, named from Stane Ford, consists of East Stamford Bridge (315) in Pocklington Rural District and West Stamford Bridge with Scoreby (115) in Escrick R.D. Ancient British or Roman road, pp. 98, 138. Battle in 1066, p. 87.

Sutton in Holderness (2397). In 1347 Sir John de Sutton made the church collegiate for 6 chaplains and so it continued until the dissolution in 1541. Nave and chancel Dec., W. tower Perp. Branceholme Castle site is in the parish, p. 127.

Thwing (307). The church has a Norm. chancel arch and S. doorway of nave, the tympanum carved with an Agnus Dei. Birthplace of John of Bridlington, p. 149, and at Octon (Roman camp, p. 100) in the parish, of Thomas Lamplugh, archbishop of York, p. 149. One-third of a mile W. of Wold Cottage in this parish a meteorite (aerolite) fell on December 13, 1795. It weighed 56 lbs., and 20,111 grams of it are in the British Museum. An obelisk commemorates the fall.

Wetwang (484). The lower part of the W. tower of the church is E.E., the three eastern bays of nave Trans., the N. transept early Dec.

Willerby (1319). The village was originally Danish, p. 138. The church has an E.E. nave arcade and Perp. W. tower.

Winestead (154), 1½ miles N.W. of Patrington. The church has no tower and is chiefly Perp., though some of the walls may be Norman and the S. aisle was rebuilt between 1889 and 1900.

The brasses and other monuments, especially of the Hildyards, are interesting. Winestead Hall, p. 136. Andrew Marvell was born in Winestead, p. 154.

Withernsea (4702), urban district, is described, p. 47. Geology, p. 31. Coast erosion, pp. 48, 50, 52. Fisheries, pp. 77, 78. Railway, p. 144.

Wold Newton (265), 2 miles N. of Thwing. Church has Norm. chancel arch and S. doorway with carved tympanum.

York (84,052), the county town of the ancient county of York-shire, a parliamentary and county borough and archiepiscopal city, is situated on both banks of the Ouse (discharge of which see p. 12). During the Roman occupation it was, as Eburacum (or Eboracum), the first or second city of Britain ; in Anglian and Danish times the capital of the kingdom of Deira; afterwards for centuries the second city of England. It lies on Ermine Street and on the Great North Road, and is an important railway centre. Its chief features have been already described, viz., the Roman roads, pp. 97, 98, and fortress, p. 98; Anglo-Saxon antiquities, p. 102; the castle, p. 130; the medieval walls and gates, p. 130; the Cathedral, p. 119; St Mary's Abbey, p. 115; other religious houses, p. 115, and churches, p. 119; Norman doorways, pp. 108, 110; half-timber houses, p. 132; street names, p. 63; river navi-gation, p. 139; railways, p. 143; history, pp. 85, 86, 87, 88, 91; industries, pp. 68, 71. The municipal corporation consists of a lord mayor, 12 aldermen and 36 councillors. In the Guildhall is a handsome old hall in Perp. style, erected in 1446. For the Merchants' Hall see p. 132, for the King's Manor House, the Treasurer's House, p. 135. The Museum of the Yorkshire Philosophical Society contains antiquities and natural history collections. There are many interesting old churches. York, a county of itself since 1396, is a quarter sessions borough. The assizes of the N. and E. Ridings and the City of York are held at York. In the Ministry of Health official list and for Board of Control and Mental Deficiency purposes, also in Agricultural Statistics, York is treated as being in the E. Riding. The Lord Lieutenant of the W. Riding is also Lord Lieutenant of York, and York is associated with that Riding for military purposes such as the Territorial Association. York returns one member to Parliament. Etymology of name, p. 2. Geology, p. 28. Climate, p. 56. Birthplace of Alcuin, p. 148. Etty, Flaxman and Barnby, p. 154. Mrs Stannard, p. 154.

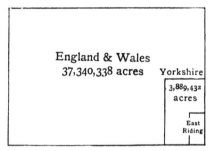

Fig. 1. Area of the East Riding (750,115
acres) compared with that of Yorkshire,
and of England and Wales

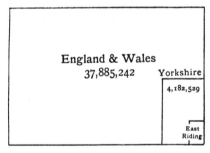

Fig. 2. Population of the East Riding
including Hull, in 1921 (460,880) compared with that of Yorkshire, and of
England and Wales

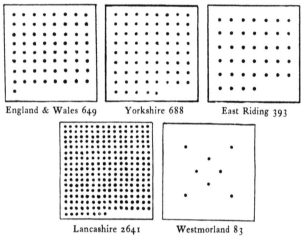

England & Wales 649   Yorkshire 688   East Riding 393

Lancashire 2641   Westmorland 83

Fig. 3. Comparative Density of the Population
to the square mile at the Census of 1921

(Each dot represents ten persons)

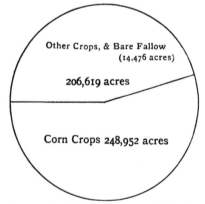

Other Crops, & Bare Fallow
(14,476 acres)

206,619 acres

Corn Crops 248,952 acres

Fig. 4. Area under Cereals compared with that of other
Cultivated Land in the East Riding in 1922

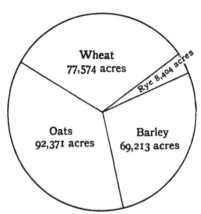

Fig. 5. Proportionate Areas of Chief Cereals in
the East Riding in 1922. In addition there were
1390 acres of Mixed Corn

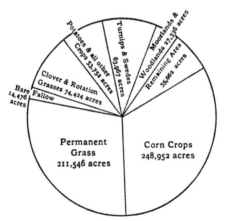

Fig. 6. Proportionate Areas of Cultivated and
Uncultivated Land in the East Riding in 1922

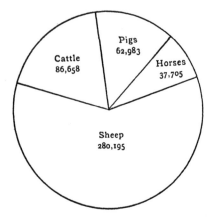

Fig. 7. Proportionate numbers of Live Stock
in the East Riding in 1922

Milton Keynes UK
Ingram Content Group UK Ltd.
UKHW032321161024
449665UK00001B/10